U.S. Department
of Transportation
**Federal Aviation
Administration**

MW01482118

Effective July 1995

Airline Transport Pilot and Type Rating

For Airplane & Helicopter

Practical Test Standards

Flight Standards Service
Washington, DC 20591

Reprinted by Aviation Supplies & Academics, Inc.
Newcastle, WA 98059-3153

FAA-S-8081-5B

**Airline Transport Pilot and Type Rating
For Airplane and Helicopter
Practical Test Standards**

Aviation Supplies & Academics, Inc.
7005 132nd Place SE
Newcastle, Washington 98059-3153

Cover design © 1995 Aviation Supplies & Academics, Inc.
All rights reserved. Published 1995.

This book is an exact reprint of the text from the *Airline Transport Pilot and/or Type Rating Practical Test Standards* by the Flight Standards Service, available on FedWorld. The publisher has made minor adjustments to the organization of the material for the sake of consistency in this printed presentation.

Printed in the United States of America

99 98 97 96 95   9 8 7 6 5 4 3 2 1

ISBN 1-56027-232-5
ASA-8081-5B

Update to Airline Transport Pilot and Type Rating Practical Test Standards

FAA-S-8081-5B   Change 2: Effective March 28, 1996

The following replaces Page 2-13 and the first paragraph on Page 2-14 for Helicopters:

IV. INFLIGHT MANEUVERS

A. TASK: STEEP TURNS

REFERENCES: FAR Part 61; AC 61-27; FSB Report; Pilot's Operating Handbook, RFM.

Objective. To determine that the applicant:

1. Exhibits adequate knowledge of steep turns (if applicable to helicopter) and the factors associated with performance; and, if applicable, angle of bank, and pitch and power requirements.
2. **Selects an altitude recommended by the manufacturer, training syllabus, or other training directive.**
3. Establishes the recommended entry airspeed.
4. Rolls into a coordinated turn of 180° or 360° with a bank as appropriate, not to exceed 30°. Maintains the bank angle within ±5° while in smooth, stabilized flight.
5. Applies smooth coordinated pitch, bank, and power to maintain the specified altitude within ±100 feet (30 meters) and the desired airspeed within ±10 knots.
6. Rolls out of the turn (at approximately the same rate as used to roll into the turn) within ±10° of the entry or specified heading, stabilizes the helicopter in a straight-and-level attitude or, at the discretion of the examiner, reverses the direction of turn and repeats the maneuver in the opposite direction.
7. Avoids any indication of abnormal flight attitude, or exceeding any structural, rotor, or operating limitation during any part of the maneuver.

B. TASK: POWERPLANT FAILURE — MULTIENGINE HELICOPTER

REFERENCES: FAR Part 61; Pilot's Operating Handbook, RFM.

NOTE: When this task is accomplished in an approved flight simulator, the engine shutdown and restart may be performed in conjunction with another procedure or maneuver, and at any location or altitude at the discretion of the examiner.

When this task is accomplished in the helicopter, the engine failure and restart procedure shall be simulated. This task shall be performed by reducing the power to idle on the selected engine. This task must be initiated at an altitude from which a safe landing can be made in the event of actual engine problems.

FOREWORD

The Airline Transport Pilot and/or Type Rating Practical Test Standards (PTS) book has been published by the Federal Aviation Administration (FAA) to establish the standards for Airline Transport Pilot and Type Rating practical tests for airplanes and helicopters. FAA inspectors, designated pilot examiners, and check airmen (referred to as examiners throughout the remaining PTS) shall conduct practical tests in compliance with these standards. Flight instructors and applicants should find these standards helpful in practical test preparation.

NOTE

FAA-S-8081-5A, Airline Transport Pilot and Type Rating Practical Test Standards, dated May 1995, was prematurely printed and canceled in April 1995. Examiners may use the practical test standards, dated August 1988, to evaluate airline transport pilot and type rating applicants until July 1, 1995. After that date, the practical test standards, dated August 1988, will be superseded and the practical test standards, dated July 1995, shall be in effect.

Thomas C. Accardi
Director, Flight Standards Service

INTRODUCTION

The Flight Standards Service of the Federal Aviation Administration (FAA) has developed this Practical Test Standard (PTS) to be used by examiners when conducting airman practical tests (knowledge of the equipment and flight tasks). Instructors are expected to use this PTS when preparing applicants for practical tests.

This publication sets forth the practical test requirements for the airline transport pilot certificate in airplanes or helicopters or the addition of an aircraft type rating in airplanes or helicopters.

This publication may be purchased from the Superintendent of Documents, U.S. Government Printing Office, Washington, DC 20402.

Comments regarding this publication should be sent to:

U.S. Department of Transportation
Federal Aviation Administration
Flight Standards Service
Operations Support Branch, AFS-630
P.O. Box 25082
Oklahoma City, Oklahoma 73125

PRACTICAL TEST STANDARD CONCEPT

Federal Aviation Regulations (FARs) specify the areas in which knowledge and skills must be demonstrated by the applicant before the issuance of an airline transport pilot certificate or type rating in airplanes or helicopters. The FARs provide the flexibility to permit the FAA to publish practical test standards containing areas of knowledge and skill identified for specific TASKS (procedure and maneuver) in which pilot competency (proficiency) must be demonstrated. The FAA will add, delete, or revise TASKS and their related knowledge and skills whenever it is determined that changes are needed in the interest of safety. Adherence to provisions of the regulations and the PTS is mandatory for the evaluation of pilot applicants. For some aircraft types, provisions of FAA Flight Standardization Board (FSB) Reports may specify details as to how the FARs and this PTS apply to certain maneuvers, TASKS, procedures or knowledge areas. Airmen certification credits applicable to FAR 121 and 135 operators may be permitted in accordance with SFAR 58, AC 120-53, and FSB Reports.

The REFERENCE identifies the publication(s) that describe(s) the TASK. Descriptions of TASKS are not included in the standards because this information can be found in the listed references, as amended. Publications other than those listed may be used for references if their content conveys substantially the same meaning as the referenced publications. This PTS is based on the following references.

FAR Part 61	Certification: Pilots and Flight Instructors
FAR Part 91	General Operating and Flight Rules
SFAR 58	Advanced Qualification Program
AC 00-6	Aviation Weather
AC 00-45	Aviation Weather Services
AC 61-13	Basic Helicopter Handbook
AC 61-21	Flight Training Handbook
AC 61-27	Instrument Flying Handbook
AC 61-84	Role of Preflight Preparation
AC 120-28	Criteria for Approval of Category III Landing Weather Minima
AC 120-29	Criteria for Approving Category I and Category II Landing Minima for FAR 121 Operators
AC 120-40	Airplane Simulator Qualification
AC 120-45	Airplane Flight Training Device Qualification
AC 120-46	Use of Advanced Training Devices (Airplane Only)
AC 120-51	Crew Resource Management Training
AC 120-53	Crew Qualification and Pilot Type Rating Requirements for Transport Category Aircraft Operated Under FAR Part 121
AC 120-54	Advanced Qualification Program
AC 120-63	Helicopter Simulator Qualification
FSB Reports	Flight Standardization Board Reports
AIM	Airman's Information Manual
	Pertinent Pilot Operating Handbooks and Flight Manuals
	En Route Low and High Altitude Charts
SID	Standard Instrument Departure
STAR	Standard Terminal Arrivals
AFD	Airport Facility Directory
FDC NOTAM	National Flight Data Center Notices to Airmen
IAP	Instrument Approach Procedure

Note: The latest revision of these references should be used.

The Objective lists, in sequence, the important ELEMENTS that must be satisfactorily performed to demonstrate competency in a TASK. The Objective includes:

1. specifically what the applicant should be able to do;
2. the conditions under which the TASK is to be performed; and
3. the minimum acceptable standards of performance.

This practical test standard (PTS) is a directive designed to provide instructions, guidance, and requirements for use by examiners while evaluating pilot applicants.

Information considered directive in nature is described in this practical test standard in terms such as "shall" and "must," and means that the actions are mandatory. Guidance information is described in terms such as "should" or "may," and indicate actions that are desirable, permissive, or not mandatory and provide for flexibility.

USE OF THE PRACTICAL TEST STANDARDS

This PTS is divided into two parts: Part 1 for Airplanes and Part 2 for Helicopters. The TASKS in Part 1 apply to airplanes and the TASKS in Part 2 apply to helicopters. These TASKS apply to the applicant who seeks an airline transport pilot certificate; the addition of a category, class, or type rating on that certificate; and to the applicant who holds a private or commercial pilot certificate and is seeking the addition of a type rating on that certificate.

With certain exceptions, some described by NOTES, all TASKS in each part are required. However, when a particular ELEMENT is not appropriate to the aircraft or its equipment, that ELEMENT, at the discretion of the examiner, may be omitted. Examples of ELEMENT exceptions are high altitude weather phenomena for helicopters, integrated flight systems for aircraft not so equipped, operation of landing gear in fixed gear aircraft, multiengine tasks in single-engine aircraft, or other situations where the aircraft operation is not compatible with the requirement of the ELEMENT.

It is not intended that the examiner follow the precise order in which the AREAS OF OPERATION and TASKS appear in the PTS. The examiner may change the sequence or combine TASKS with similar Objectives to conserve time. Examiners must develop a written plan of action that includes the order and combination of TASKS to be demonstrated by the applicant in a manner that results in an efficient and valid test. Although TASKS with similar Objectives may be combined to conserve time, the Objectives of all TASKS must be demonstrated and evaluated at some time during the practical test. It is of utmost importance that the examiner accurately evaluate the applicant's ability to perform safely as a pilot in the National Airspace System. The examiner may act as air traffic control (ATC) while conducting the practical test.

Examiners shall place special emphasis upon areas of aircraft operations considered critical to flight safety. Among these are positive aircraft control, positive exchange of the flight controls procedure (who is flying the aircraft), collision avoidance, wake turbulence avoidance, use of available automation, communication management, crew resource management, and other areas deemed appropriate to any phase of the practical test. Although these areas may not be specifically addressed under each TASK, they are essential to flight safety and will be critically evaluated during the practical test. In all instances, the applicant's actions will relate to the complete situation. The examiner's role regarding ATC, crew resource management, and the duties and responsibilities of the examiner through all phases of the practical test must be explained to and understood by the applicant.

METRIC CONVERSION INITIATIVE

To assist the pilots in understanding and using the metric measurement system, the PTS refer to the metric equivalent of various altitudes throughout. The inclusion of meters is intended to familiarize pilots with its use. The metric altimeter is arranged in 10 meter increments; therefore, when converting from feet to meters, the exact conversion, being too exact for practical purposes, is rounded to the nearest 10 meter increment.

EXAMINER RESPONSIBILITY

The examiner who conducts the practical test is responsible for determining that the applicant meets the standards outlined in the Objective of each TASK within the AREAS OF OPERATION, in the PTS. The examiner shall meet this responsibility by determining that the applicant's knowledge and skill meets the Objective in all required TASKS.

Each part of this practical test standard has AREAS OF OPERATION divided into two sections. The first section is conducted on the ground to determine the applicant's knowledge of the aircraft equipment, performance and limitations. The second section consists of nine AREAS OF OPERATION; the first of which is conducted on the ground and the remaining eight are considered to be inflight. All nine AREAS OF OPERATION in the second section test the applicant's skill and knowledge.

The equipment examination must be closely coordinated and related to the flight portion of the practical test but must not be given during the flight portion of the practical test. The equipment examination should be administered prior (it may be the same day) to the flight portion of the practical test.

The examiner may accept written evidence of the equipment exam if the exam is approved by the Administrator and administered by an individual authorized by the Administrator. The examiner shall use whatever means deemed suitable to determine that the applicant's equipment knowledge meets standard.

The AREAS OF OPERATION in Section 2 contain TASKS which include both "knowledge" and "skill" ELEMENTS. The examiner shall ask the applicant to perform the skill ELEMENTS. Knowledge ELEMENTS not evident in the demonstrated skills may be tested by questioning, at anytime, during the flight event. Questioning in flight should be used judiciously so that safety is not jeopardized. Questions may be deferred until after the flight test is completed.

For aircraft requiring only one pilot, the examiner may not assist the applicant in the management of the aircraft, radio communications, tuning and identifying navigational equipment, and using navigation charts. If the examiner, other than an FAA Inspector, is qualified and current in the specific make and model aircraft, that is certified for two or more crewmembers, he or she may occupy a duty position. If the examiner occupies a duty position on an aircraft that requires two or more crewmembers, the examiner must fulfill the duties of that position. Moreover, when occupying a required duty position, the examiner shall perform crew resource management functions as briefed and requested by the applicant.

Helicopters, not certified for IMC conditions, may be operated on an IFR flight plan in VMC conditions for the purpose of conducting a practical test. When conducting the practical test in a helicopter without an autopilot, or automatic stabilization equipment (ASE) examiners may act as an autopilot (e.g., hold heading and altitude) when requested, to allow the applicant to tune radios, select charts, etc. Examiners SHALL NOT perform the functions of an autopilot when entering holding, initiating or during a standard instrument approach procedure and the missed approach procedure. Examiners may perform the same functions as an autopilot but shall not act as a copilot performing more extensive duties.

SAFETY of FLIGHT shall be the prime consideration at all times. The examiner, applicant, and crew shall be constantly alert for other traffic.

CREW RESOURCE MANAGEMENT (CRM)

CRM "...refers to the effective use of all available resources; human resources, hardware, and information." Human resources "...includes all other groups routinely working with the cockpit crew (or pilot) who are involved in decisions that are required to operate a flight safely. These groups include, but are not limited to: dispatchers, cabin crewmembers, maintenance personnel, and air traffic controllers." CRM is not a single TASK. CRM is a set of skill competencies which must be evident in all TASKS in this PTS as applied to the single pilot or the multicrew operation. CRM competencies, grouped into three clusters of observable behavior, are:

1. COMMUNICATIONS PROCESSES AND DECISIONS

    a.   Briefing
    b.   Inquiry/Advocacy/Assertiveness
    c.   Self-Critique
    d.   Communication with available personnel resources
    e.   Decision making

2. BUILDING AND MAINTENANCE OF A FLIGHT TEAM

    a.   Leadership/Followership
    b.   Interpersonal Relationships

3. WORKLOAD MANAGEMENT AND SITUATIONAL AWARENESS

    a.   Preparation/Planning
    b.   Vigilance
    c.   Workload Distribution
    d.   Distraction Avoidance
    e.   Wake Turbulence Avoidance

CRM deficiencies almost always contribute to the unsatisfactory performance of a TASK. Therefore, the competencies provide an extremely valuable vocabulary for debriefing. For debriefing purposes, an amplified list of these competencies, expressed as behavioral markers, may be found in AC 120-51, as amended. These markers consider the use of various levels of automation in flight management systems.

The standards for each CRM competency as generally stated and applied are subjective. Conversely, some of the competencies may be found objectively stated as required operational procedures for one or more TASKS. Examples of the latter include briefings, radio calls, and instrument approach callouts. Whether subjective or objective, application of CRM competencies are dependent upon the composition of the crew.

HOW THE EXAMINER APPLIES CRM

Examiners are required to exercise proper CRM competencies in conducting tests as well as expecting the same from applicants.

Pass/Fail judgments based solely on CRM issues must be carefully chosen since they may be entirely subjective. Those Pass/Fail judgments which are not subjective apply to CRM-related procedures in FAA-approved operations manuals that must be accomplished, such as briefings to other crewmembers. In such cases, the operator (or the aircraft manufacturer) specifies what should be briefed and when the briefings should occur. The examiner may judge objectively whether the briefing requirement was or was not met. In those cases where the operator (or aircraft manufacturer) has not specified a briefing, the examiner shall require the applicant to brief the appropriate items from the following note. The examiner may then judge objectively whether the briefing requirement was or was not met.

NOTE: The majority of aviation accidents and incidents are due to resource management failures by the pilot/crew; fewer are due to technical failures. Each applicant shall give a crew briefing before each takeoff/departure and approach/landing. If the operator or aircraft manufacturer has not specified a briefing, the briefing shall cover the appropriate items, such as runway, SID/STAR/IAP, power settings, speeds, abnormals or emergency prior to or after V_1, emergency return intentions, missed approach procedures, FAF, altitude at FAF, initial rate of descent, DH/MDA, time to missed approach, and what is expected of the other crewmembers during the takeoff/SID and approach/landing. If the first takeoff/departure and approach/landing briefings are satisfactory, the examiner may allow the applicant to brief only the changes, during the remainder of the flight.

FAA-S-8081-5B

PRACTICAL TEST PREREQUISITES: AIRLINE TRANSPORT PILOT

An applicant for the original issuance of an airline transport pilot certificate – airplane or helicopter – is required (prior to the practical test) by FAR Part 61 to have:

1. passed the appropriate Airline Transport Pilot knowledge test within 24 months before the date of the practical test,
2. received the applicable instruction and aeronautical experience prescribed in FAR Part 61, and
3. a first-class medical certificate issued within the past 6 months.

NOTE: The 24-month limitation does not apply if —

1. The applicant —

    a. within the period ending 24 calendar months after the month in which the applicant passed the first of any required knowledge tests, was employed by a U.S. air carrier or commercial operator operating either under Part 121 or a commuter air carrier under Part 135 (as defined in Part 298 of Title 14 of the Code of Federal Regulations) and is employed by such a certificate holder at the time of the flight test;
    b. has completed initial training, and if appropriate, transition or upgrade training; and
    c. meets the recurrent training requirements of the applicable Part; or

2. within the period ending 24 calendar months after the month in which the applicant passed the first of any required knowledge tests, the applicant participated as a pilot in a pilot training program of a U.S. scheduled military air transportation service and is currently participating in that program.

PRACTICAL TEST PREREQUISITES: TYPE RATING

An applicant for a type rating in an airplane or helicopter is required by FAR Part 61 to have:

1. the applicable experience, and
2. an appropriate and valid medical certificate.

In addition, the applicant who is applying for an aircraft type rating to be added to an airline transport pilot or an aircraft type rating associated with an airline transport pilot certificate must have:

1. received and logged ground training from an authorized ground or flight instructor and flight training from an authorized flight instructor, on the approved AREAS OF OPERATION in this PTS that apply to the aircraft type rating sought; and
2. received a logbook endorsement from the instructor who conducted the training, certifying that the applicant completed all the training on the AREAS OF OPERATION in this PTS that apply to the aircraft type rating sought; or
3. if the applicant is an employee of a Part 121 or Part 135 certificate holder, the applicant may present a training record that shows the satisfactory completion of that certificate holder's approved pilot-in-command training program for the aircraft type rating sought, instead of the requirements of 1 and 2 above.

An applicant who holds the private pilot or limited commercial pilot certificate is required to have passed the appropriate instrument rating knowledge test since the beginning of the 24th month before the practical test is taken if the test is for the concurrent issuance of an instrument rating and an aircraft type rating.

If an applicant is taking a practical test for the issuance of a private or commercial pilot certificate with an airplane/helicopter rating, in an aircraft that requires a type rating, Private Pilot PTS or Commercial Pilot PTS, as appropriate to the certificate, should be used in conjunction with this guide. Also, the current Instrument Rating Practical Test Standard should be used in conjunction with this guide if the applicant is concurrently taking a practical test for the issuance of an instrument rating and a type rating. The TASKS that are in the Private Pilot, Commercial Pilot or Instrument Rating PTS's (and not in this PTS) must be accomplished.

TYPE RATINGS LIMITED TO VFR

AIRPLANES:

Pilot applicants who wish to add a type rating, limited to VFR, to their certificate must take a practical test that includes the following items, as listed on pages 1-i and 1-ii in Part 1 of this document:

Section One: AREA OF OPERATION

PREFLIGHT PREPARATION

    I. Equipment knowledge.

        A. Equipment examination.
        B. Performance and limitations.

Section Two: AREAS OF OPERATION

    I. Preflight procedures.

        A. Preflight inspection.

    II. Ground operations.

        A. Powerplant start.
        B. Taxiing.
        C. Pretakeoff checks.

    III. Takeoff and departure maneuvers.

        A. Normal and crosswind takeoff.
        B. Powerplant failure.
        C. Rejected takeoff.

    IV. Inflight maneuvers.

        A. Steep turns.
        B. Approaches to stalls.
        C. Powerplant failure — multiengine airplanes.
        D. Powerplant failure — single-engine airplanes.
        E. Specific flight characteristics.

V. Instrument procedures (Not applicable).

VI. Landings.

    A. Normal and crosswind landings.
    B. Landing with simulated powerplant
        failure — multiengine airplanes.
    C. Rejected landing.
    D. Landing with a zero or nonstandard flap setting.

VII. Normal and abnormal procedures.

VIII. Emergency procedures.

IX. Postflight procedures.

    A. After-landing.
    B. Parking and securing.

HELICOPTERS:

Pilot applicants who wish to add a type rating, limited to VFR, to their certificate must take a practical test that includes the following items, as listed on pages 2-i and 2-ii in Part 2 of this document:

Section One: AREA OF OPERATION

PREFLIGHT PREPARATION

    I. Equipment knowledge.

        A. Equipment examination.
        B. Performance and limitations.

Section Two: AREAS OF OPERATION

    I. Preflight procedures.

        A. Preflight inspection.

    II. Ground operations.

        A. Powerplant start.
        B. Taxiing.
        C. Pretakeoff checks.

III. Takeoff and departure maneuvers.

    A. Normal and crosswind takeoff.
    B. Powerplant failure.
    C. Rejected takeoff.

IV. Inflight maneuvers.

    A. Steep turns.
    B. Powerplant failure — multiengine.
    C. Powerplant failure — single-engine.
    D. Recovery from unusual attitudes.
    E. Settling with power.

V. Instrument procedures (Not applicable).

VI. Landings.

    A. Normal and crosswind landings.
    B. Landing with simulated powerplant failure — multiengine.
    C. Rejected landing.

VII. Normal and abnormal procedures.

VIII. Emergency procedures.

IX. Postflight procedures.

    A. After-landing.
    B. Parking and securing.

AIRCRAFT AND EQUIPMENT REQUIREMENTS FOR THE PRACTICAL TEST

The applicant is required to provide an appropriate and airworthy aircraft for the practical test. Its operating limitations must not prohibit the TASKS required on the practical test. Flight instruments are those required for controlling the aircraft without outside references. The aircraft must have radio equipment for communications with air traffic control and the performance of instrument approach procedures.

USE OF FLIGHT SIMULATOR OR FLIGHT TRAINING DEVICE

In the AREA OF OPERATION labeled "Preflight Preparation," the TASKS are knowledge only. These TASKS do not require the use of a flight training device (FTD), flight simulator, or an aircraft to accomplish, but any of them may be used. Each inflight maneuver or procedure must be performed by the pilot applicant in an FTD, flight simulator, or an aircraft. Appendix 1 or Appendix 2, as applicable, of this PTS should be consulted to identify the maneuvers or procedures that may be accomplished in an FTD or flight simulator. The level of FTD or flight simulator required for each maneuver or procedure will also be found in the appropriate appendix.

When accomplished in an aircraft, certain task elements may be accomplished through "simulated" actions in the interest of safety and practicality, but when accomplished in an FTD or flight simulator these same actions would not be "simulated." For example, when in an aircraft, a simulated engine fire may be addressed by retarding the throttle to idle, simulating the shutdown of the engine, simulating the discharge of the fire suppression agent, simulating the disconnection of associated electrics, hydraulics, pneumatics, etc. However, when the same emergency condition is addressed in an FTD or a flight simulator, all task elements must be accomplished as would be expected under actual circumstances. Similarly, safety of flight precautions taken in the aircraft for the accomplishment of a specific maneuver or procedure (such as limiting altitude in an approach to stall, setting maximum airspeed for an engine failure expected to result in a rejected takeoff, or limiting autorotative descents to power recoveries) need not be taken when an FTD or a flight simulator is used.

It is important to understand that whether accomplished in an FTD, a flight simulator, or the aircraft, all tasks and task elements for each maneuver or procedure will have the same performance criteria applied equally for determination of overall satisfactory performance.

SATISFACTORY PERFORMANCE

The ability of an applicant to perform the required TASKS is based on:

1. executing TASKS within the aircraft's performance capabilities and limitations, including use of the aircraft's systems;
2. executing normal, abnormal, and emergency procedures and TASKS appropriate to the aircraft;
3. piloting the aircraft with smoothness and accuracy;
4. crew resource management;
5. applying aeronautical knowledge; and
6. showing mastery of the aircraft within the standards outlined in this PTS with the successful outcome of a TASK never in doubt.

FAA-S-8081-5B

UNSATISFACTORY PERFORMANCE

Consistently exceeding tolerances stated in the TASK Objective, or failure to take prompt, corrective action when tolerances are exceeded, are indicative of unsatisfactory performance. The tolerances represent the performance expected in good flying conditions. Any action, or lack thereof, by the applicant which requires corrective intervention by the examiner to maintain safe flight shall be disqualifying.

NOTE: It is vitally important that the applicant, safety pilot, and examiner use proper and effective scanning techniques to observe all other traffic in the area to ensure the area is clear before performing any maneuvers.

When, in the judgment of the examiner, the applicant's performance of any TASK is unsatisfactory, the associated AREA OF OPERATION is failed and therefore the practical test is failed. Examiners shall not repeat TASKS that have been attempted and failed. The examiner or applicant may discontinue the test at any time after the failure of a TASK which makes the applicant ineligible for the certificate or rating sought. The practical test will be continued only with the consent of the applicant. In such cases, it is usually better for the examiner to continue with the practical test to complete the other TASKS. If the examiner determines that the entire practical test must be repeated, the practical test should not be continued but should be terminated immediately. If the practical test is either continued or discontinued, the applicant is entitled to credit for those TASKS satisfactorily performed. However, during a retest and at the discretion of the examiner, any TASK may be reevaluated including those previously passed. Whether the remaining parts of the practical test are continued or not after a failure, a notice of disapproval must be issued.

When the examiner determines that a TASK is incomplete, or the outcome uncertain, the examiner may require the applicant to repeat that TASK, or portions of that TASK. This provision has been made in the interest of fairness and does not mean that instruction or practice is permitted during the certification process. When practical, the remaining TASKS of the practical test phase should be completed before repeating the questionable TASK. If the second attempt to perform a questionable TASK is not clearly satisfactory, the examiner shall consider it unsatisfactory.

If the practical test must be terminated for unsatisfactory performance and there are other TASKS which have not been tested or still need to be repeated, a notice of disapproval shall be issued listing the specific TASKS which have not been successfully completed or tested.

When a practical test is discontinued for reasons other than unsatisfactory performance (i.e., equipment failure, weather, air sickness), FAA Form 8710-1, Airman Certificate and/or Rating Application, and, if applicable, AC Form 8080-2, Airman Written Test Report, should be returned to the applicant. The examiner at that time should prepare, sign, and issue a Letter of Discontinuance to the applicant. The Letter of Discontinuance should identify the portions of the practical test that were successfully completed.

RECORDING UNSATISFACTORY PERFORMANCE

This PTS uses the terms "AREA OF OPERATION" and "TASK" to denote areas in which competency must be demonstrated. When a disapproval notice is issued, the examiner must record the applicant's unsatisfactory performance in terms of AREA OF OPERATION and TASK appropriate to the practical test conducted.

Note: Under certain conditions, some TASKS specified in an operator's approved flight training program may be waived. However, this TASK waiver only applies to pilots who are employed by a Part 121 certificate holder and who are seeking an airline transport pilot certificate with associated airplane class and type ratings. For specific TASK waiver authority, refer to FAR Section 61.157(c).

xx

Part 1 - Airplanes

CONTENTS

PART 1, AIRPLANES
SECTION ONE: AREA OF OPERATION

PREFLIGHT PREPARATION

I. EQUIPMENT KNOWLEDGE

A. TASK: EQUIPMENT EXAMINATION

REFERENCES: FAR Part 61; Pilot's Operating Handbook, FAA Approved Airplane Flight Manual (AFM).

Objective. To determine that the applicant:

1. Exhibits adequate knowledge appropriate to the airplane; its systems and components; its normal, abnormal, and emergency procedures; and uses the correct terminology with regard to the following items —

   a. landing gear — indicators, brakes, antiskid, tires, nose-wheel steering, and shock absorbers.
   b. Powerplant — controls and indications, induction system, carburetor and fuel injection, turbocharging, cooling, fire detection/protection, mounting points, turbine wheels, compressors, deicing, anti-icing, and other related components.
   c. Propellers — type, controls, feathering/unfeathering, autofeather, negative torque sensing, synchronizing, and synchrophasing.
   d. fuel system — capacity; drains; pumps; controls; indicators; crossfeeding; transferring; jettison; fuel grade, color and additives; fueling and defueling procedures; and substitutions, if applicable.
   e. oil system — capacity, grade, quantities, and indicators.
   f. hydraulic system — capacity, pumps, pressure, reservoirs, grade, and regulators.
   g. electrical system — alternators, generators, battery, circuit breakers and protection devices, controls, indicators, and external and auxiliary power sources and ratings.
   h. environmental systems — heating, cooling, ventilation, oxygen and pressurization, controls, indicators, and regulating devices.

i.　avionics and communications — autopilot; flight director; Electronic Flight Indicating Systems (EFIS); Flight Management System(s) (FMS); Long Range Navigation (LORAN) systems; Doppler Radar; Inertial Navigation Systems (INS); Global Positioning System (GPS/DGPS-/WGPS); VOR, NDB, ILS/MLS, RNAV systems and components; indicating devices; transponder; and emergency locator transmitter.

j.　ice protection — anti-ice, deice, pitot-static system protection, propeller, windshield, wing and tail surfaces.

k.　crewmember and passenger equipment — oxygen system, survival gear, emergency exits, evacuation procedures and crew duties, and quick donning oxygen mask for crewmembers and passengers.

l.　flight controls — ailerons, elevator(s), rudder(s), control tabs, balance tabs, stabilizer, flaps, spoilers, leading edge flaps/slats and trim systems.

2.　Exhibits adequate knowledge of the contents of the Operating Handbook or AFM with regard to the systems and components listed in paragraph 1 (above); the Minimum Equipment List (MEL), if appropriate, and the Operations Specifications, if applicable.

B. TASK: PERFORMANCE AND LIMITATIONS

REFERENCES: FAR Parts 1, 61, 91; Pilot's Operating Handbook, AFM.

Objective. To determine that the applicant:

1.　Exhibits adequate knowledge of performance and limitations, including a thorough knowledge of the adverse effects of exceeding any limitation.

2.　Demonstrates proficient use of (as appropriate to the airplane) performance charts, tables, graphs, or other data relating to items such as —

a.　accelerate-stop distance.

b.　accelerate-go distance.

c.　takeoff performance, all engines, engine(s) inoperative.

d.　climb performance including segmented climb performance; with all engines operating; with one or more engine(s) inoperative, and with other engine malfunctions as may be appropriate.

    e.   service ceiling, all engines, engines(s) inoperative, including Drift Down, if appropriate.
    f.   cruise performance.
    g.   fuel consumption, range, and endurance.
    h.   descent performance.
    i.   go-around from rejected landings.
    j.   other performance data (appropriate to the airplane).

3. Describes (as appropriate to the airplane) the airspeeds used during specific phases of flight.

4. Describes the effects of meteorological conditions upon performance characteristics and correctly applies these factors to a specific chart, table, graph or other performance data.

5. Computes the center-of-gravity location for a specific load condition (as specified by the examiner), including adding, removing, or shifting weight.

6. Determines if the computed center of gravity is within the forward and aft center-of-gravity limits, and that lateral fuel balance is within limits for takeoff and landing.

7. Demonstrates good planning and knowledge of procedures in applying operational factors affecting airplane performance.

FAA-S-8081-5B

PART 1, AIRPLANES
SECTION TWO: AREAS OF OPERATION

I. PREFLIGHT PROCEDURES

A. TASK: PREFLIGHT INSPECTION

REFERENCES: FAR Parts 61, 91; Pilot's Operating Handbook, AFM.

NOTE: If a flight engineer (FE) is a required crewmember for a particular type airplane, the actual visual inspection may be waived. The actual visual inspection may be replaced by using an approved pictorial means that realistically portrays the location and detail of inspection items. On airplanes requiring an FE, an applicant must demonstrate adequate knowledge of the FE functions for the safe completion of the flight if the FE becomes ill or incapacitated during a flight.

Objective. To determine that the applicant:

1. Exhibits adequate knowledge of the preflight inspection procedures, while explaining briefly —

    a. the purpose of inspecting the items which must be checked.
    b. how to detect possible defects.
    c. the corrective action to take.

2. Exhibits adequate knowledge of the operational status of the airplane by locating and explaining the significance and importance of related documents such as —

    a. airworthiness and registration certificates.
    b. operating limitations, handbooks, and manuals.
    c. minimum equipment list (MEL) (if appropriate).
    d. weight and balance data.
    e. maintenance requirements, tests, and appropriate records applicable to the proposed flight or operation; and maintenance that may be performed by the pilot or other designated crewmember.

3. Uses the approved checklist to inspect the airplane externally and internally.

4. Uses the challenge-and-response (or other approved) method with the other crewmember(s), where applicable, to accomplish the checklist procedures.
5. Verifies the airplane is safe for flight by emphasizing (as appropriate) the need to look at and explain the purpose of inspecting items such as —

    a. powerplant, including controls and indicators.
    b. fuel quantity, grade, type, contamination safeguards, and servicing procedures.
    c. oil quantity, grade, and type.
    d. hydraulic fluid quantity, grade, type, and servicing procedures.
    e. oxygen quantity, pressures, servicing procedures, and associated systems and equipment for crew and passengers.
    f. landing gear, brakes, and steering system.
    g. tires for condition, inflation, and correct mounting, where applicable.
    h. fire protection/detection systems for proper operation, servicing, pressures, and discharge indications.
    i. pneumatic system pressures and servicing.
    j. ground environmental systems for proper servicing and operation.
    k. auxiliary power unit (APU) for servicing and operation.
    l. flight control systems including trim, spoilers, and leading/trailing edge.
    m. anti-ice, deice systems, servicing, and operation.

6. Coordinates with ground crew and ensures adequate clearance prior to moving any devices such as door, hatches, and flight control surfaces.
7. Complies with the provisions of the appropriate Operations Specifications, if applicable, as they pertain to the particular airplane and operation.
8. Demonstrates proper operation of all applicable airplane systems.
9. Notes any discrepancies, determines if the airplane is airworthy and safe for flight, or takes the proper corrective action.
10. Checks the general area around the airplane for hazards to the safety of the airplane and personnel.

II. GROUND OPERATIONS

A. TASK: POWERPLANT START

REFERENCES: FAR Part 61; Pilot's Operating Handbook, AFM.

Objective. To determine that the applicant:

1. Exhibits adequate knowledge of the correct powerplant start procedures including the use of an auxiliary power unit (APU) or external power source, starting under various atmospheric conditions, normal and abnormal starting limitations, and the proper action required in the event of a malfunction.
2. Ensures the ground safety procedures are followed during the before-start, start, and after-start phases.
3. Ensures the use of appropriate ground crew personnel during the start procedures.
4. Performs all items of the start procedures by systematically following the approved checklist items for the before-start, start, and after-start phases.
5. Demonstrates sound judgment and operating practices in those instances where specific instructions or checklist items are not published.

B. TASK: TAXIING

REFERENCES: FAR Part 61; Pilot's Operating Handbook, AFM.

Objective. To determine that the applicant:

1. Exhibits adequate knowledge of safe taxi procedures (as appropriate to the airplane including push-back or power-back, as may be applicable).
2. Demonstrates proficiency by maintaining correct and positive airplane control. In airplanes equipped with float devices, this includes water taxiing, approaching a buoy, and docking.
3. Maintains proper spacing on other aircraft, obstructions, and persons.
4. Accomplishes the applicable checklist items and performs recommended procedures.
5. Maintains desired track and speed.
6. Complies with instructions issued by ATC (or the examiner simulating ATC).

7. Observes runway hold lines, localizer and glide slope critical areas, and other surface control markings and lighting.
8. Maintains constant vigilance and airplane control during taxi operation.

C. TASK: PRETAKEOFF CHECKS

REFERENCES: FAR Part 61; Pilot's Operating Handbook, AFM.

Objective. To determine that the applicant:

1. Exhibits adequate knowledge of the pretakeoff checks by stating the reason for checking the items outlined on the approved checklist and explaining how to detect possible malfunctions.
2. Divides attention properly inside and outside cockpit.
3. Ensures that all systems are within their normal operating range prior to beginning, during the performance of, and at the completion of those checks required by the approved checklist.
4. Explains, as may be requested by the examiner, any normal or abnormal system operating characteristic or limitation; and the corrective action for a specific malfunction.
5. Determines if the airplane is safe for the proposed flight or requires maintenance.
6. Determines the airplane's takeoff performance, considering such factors as wind, density altitude, weight, temperature, pressure altitude, and runway condition and length.
7. Determines airspeeds/V-speeds and properly sets all instrument references, flight director and autopilot controls, and navigation and communications equipment.
8. Reviews procedures for emergency and abnormal situations which may be encountered during takeoff, and states the corrective action required of the pilot in command and other concerned crewmembers.
9. Obtains and correctly interprets the takeoff and departure clearance as issued by ATC.

III. TAKEOFF AND DEPARTURE MANEUVERS

A. TASK: NORMAL AND CROSSWIND TAKEOFF

REFERENCES: FAR Part 61; Pilot's Operating Handbook, AFM.

Objective. To determine that the applicant:

1. Exhibits adequate knowledge of normal and crosswind takeoffs and climbs including (as appropriate to the airplane) airspeeds, configurations, and emergency/ abnormal procedures.
2. Notes any obstructions or other hazards that might hinder a safe takeoff.
3. Verifies and correctly applies correction for the existing wind component to the takeoff performance.
4. Completes required checks prior to starting takeoff to verify the expected powerplant performance. Performs all required pretakeoff checks as required by the appropriate checklist items.
5. Aligns the airplane on the runway centerline.
6. Applies the controls correctly to maintain longitudinal alignment on the centerline of the runway prior to initiating and during the takeoff.
7. Adjusts the powerplant controls as recommended by the FAA-approved guidance for the existing conditions.
8. Monitors powerplant controls, settings, and instruments during takeoff to ensure all predetermined parameters are maintained.
9. Adjusts the controls to attain the desired pitch attitude at the predetermined airspeed/V-speed to attain the desired performance for the particular takeoff segment.
10. Performs the required pitch changes and, as appropriate, performs or calls for and verifies the accomplishment of, gear and flap retractions, power adjustments, and other required pilot-related activities at the required airspeed/V-speeds within the tolerances established in the Pilot's Operating Handbook or AFM.
11. Uses the applicable noise abatement, wake turbulence avoidance procedures, as required.
12. Accomplishes or calls for and verifies the accomplishment of the appropriate checklist items.
13. Maintains the appropriate climb segment airspeed/V-speeds.
14. Maintains the desired heading within ±5° and the desired airspeed/V-speed within ±5 knots or the appropriate V-speed range.

B. TASK: INSTRUMENT TAKEOFF

REFERENCES: FAR Part 61; AC 61-27; Pilot's Operating Handbook, AFM, AIM.

Objective. To determine that the applicant:

1. Exhibits adequate knowledge of an instrument takeoff with instrument meteorological conditions simulated at or before reaching an altitude of 100 feet (30 meters) AGL. If accomplished in a flight simulator, visibility should be no greater than one-quarter (1/4) mile, or as specified by operator specifications.
2. Takes into account, prior to beginning the takeoff, operational factors which could affect the maneuver such as Takeoff Warning Inhibit Systems or other airplane characteristics, runway length, surface conditions, wind, wake turbulence, obstructions, and other related factors that could adversely affect safety.
3. Accomplishes the appropriate checklist items to ensure that the airplane systems applicable to the instrument takeoff are operating properly.
4. Sets the applicable radios/flight instruments to the desired setting prior to initiating the takeoff.
5. Applies the controls correctly to maintain longitudinal alignment on the centerline of the runway prior to initiating and during the takeoff.
6. Transitions smoothly and accurately from visual meteorological conditions to actual or simulated instrument meteorological conditions.
7. Maintains the appropriate climb attitude.
8. Complies with the appropriate airspeeds/V-speeds and climb segment airspeeds.
9. Maintains desired heading within ±5° and desired airspeeds within ±5 knots.
10. Complies with ATC clearances and instructions issued by ATC (or the examiner simulating ATC).

C. TASK: POWERPLANT FAILURE DURING TAKEOFF

REFERENCES: FAR Part 61; AC 61-21; Pilot's Operating Handbook, AFM; DOT/FAA Takeoff Safety Training Aid.

Objective. To determine that the applicant:

1. Exhibits adequate knowledge of the procedures used during powerplant failure on takeoff, the appropriate reference airspeeds, and the specific pilot actions required.
2. Takes into account, prior to beginning the takeoff, operational factors which could affect the maneuver such as Takeoff Warning Inhibit Systems or other airplane characteristics, runway length, surface conditions, wind, wake turbulence, obstructions, and other related factors that could adversely affect safety.
3. Completes required checks prior to starting takeoff to verify the expected powerplant performance. Performs all required pretakeoff checks as required by the appropriate checklist items.
4. Aligns the airplane on the runway.
5. Applies the controls correctly to maintain longitudinal alignment on the centerline of the runway prior to initiating and during the takeoff.
6. Adjusts the powerplant controls as recommended by the FAA-approved guidance for the existing conditions.
7. Single-Engine Airplanes: Establishes a power-off descent approximately straight-ahead, if the powerplant failure occurs after becoming airborne.
8. Continues the takeoff (in a multiengine airplane) if the powerplant failure occurs at a point where the airplane can continue to a specified airspeed and altitude at the end of the runway commensurate with the airplane's performance capabilities and operating limitations.
9. Maintains (in a multiengine airplane), after a simulated powerplant failure and after a climb has been established, the desired heading within ±5°, desired airspeed within ±5 knots, and, if appropriate for the airplane, establishes a bank of approximately 5°, or as recommended by the manufacturer, toward the operating powerplant.
10. In a multiengine airplane with published V_1, V_R, and/or V_2 speeds, the failure of the most critical powerplant should be simulated at a point:

    a. After V_1 and prior to V_2, if in the opinion of the examiner, it is appropriate under the prevailing conditions; or

b. As close as possible after V_1 when V_1 and V_2 or V_1 and V_R are identical.

11. In a multiengine airplane for which no V_1, V_R, or V_2 speeds are published, the failure of the most critical powerplant should be simulated at a point after reaching a minimum of V_{MCA} and, if accomplished in the aircraft, at an altitude not lower than 500 feet AGL.

12. Maintains the airplane alignment with the heading appropriate for climb performance and terrain clearance when powerplant failure occurs.

D. TASK: REJECTED TAKEOFF

REFERENCES: FAR Part 61; AC 61-21; Pilot's Operating Handbook, AFM; DOT/FAA Takeoff Safety Training Aid.

Objective. To determine that the applicant understands when to reject or continue the takeoff:

1. Exhibits adequate knowledge of the technique and procedure for accomplishing a rejected takeoff after powerplant/system(s) failure/warnings, including related safety factors.

2. Takes into account, prior to beginning the takeoff, operational factors which could affect the maneuver such as Takeoff Warning Inhibit Systems or other airplane characteristics, runway length, surface conditions, wind, obstructions, and other related factors that could affect takeoff performance and could adversely affect safety.

3. Aligns the airplane on the runway centerline.

4. Performs all required pretakeoff checks as required by the appropriate checklist items.

5. Adjusts the powerplant controls as recommended by the FAA-approved guidance for the existing conditions.

6. Applies the controls correctly to maintain longitudinal alignment on the centerline of the runway.

7. Aborts the takeoff if, in a single-engine airplane the powerplant failure occurs prior to becoming airborne, or in a multiengine airplane, the powerplant failure occurs at a point during the takeoff where the abort procedure can be initiated and the airplane can be safely stopped on the remaining runway/stopway. If a flight simulator is not used, the powerplant failure should be simulated before reaching 50 percent of V_{MC}.

8. Reduces the power smoothly and promptly, if appropriate to the airplane, when powerplant failure is recognized.

9. Uses spoilers, prop reverse, thrust reverse, wheel brakes, and other drag/braking devices, as appropriate, maintaining positive control in such a manner as to bring the airplane to a safe stop. Accomplishes the appropriate powerplant failure or other procedures and/or checklists as set forth in the pilot operating hand-book or AFM.

E. TASK: INSTRUMENT DEPARTURE

REFERENCES: FAR Part 61; AC 61-27; Pilot's Operating Handbook, AFM, AIM.

Objective. To determine that the applicant:

1. Exhibits adequate knowledge of SIDs, En Route Low/High Altitude Charts, STARs and related pilot/controller responsibilities.
2. Uses the current and appropriate navigation publications for the proposed flight.
3. Selects and uses the appropriate communications frequencies, and selects and identifies the navigation aids associated with the proposed flight.
4. Performs the appropriate checklist items.
5. Establishes communications with ATC, using proper phraseology.
6. Complies, in a timely manner, with all instructions and airspace restrictions.
7. Exhibits adequate knowledge of two-way radio communications failure procedures.
8. Intercepts, in a timely manner, all courses, radials, and bearings appropriate to the procedure, route, clearance, or as directed by the examiner.
9. Maintains the appropriate airspeed within ±10 knots, headings within ±10°, altitude within ±100 feet (30 meters); and accurately tracks a course, radial, or bearing.
10. Conducts the departure phase to a point where, in the opinion of the examiner, the transition to the en route environment is complete.

IV. INFLIGHT MANEUVERS

A. TASK: STEEP TURNS

REFERENCES: FAR Part 61; AC 61-27; FSB Report; Pilot's Operating Handbook, AFM.

Objective. To determine that the applicant:

1. Exhibits adequate knowledge of steep turns (if applicable to the airplane) and the factors associated with performance; and, if applicable, wing loading, angle of bank, stall speed, pitch, power requirements, and over-banking tendencies.
2. Selects an altitude recommended by the manufacturer, training syllabus, or other training directive, but in no case lower than 3,000 feet (900 meters) AGL.
3. Establishes the recommended entry airspeed.
4. Rolls into a coordinated turn of 180° or 360° with a bank of at least 45°. Maintains the bank angle within ±5° while in smooth, stabilized flight.
5. Applies smooth coordinated pitch, bank, and power to maintain the specified altitude within ±100 feet (30 meters) and the desired airspeed within ±10 knots.
6. Rolls out of the turn (at approximately the same rate as used to roll into the turn) within ±10° of the entry or specified heading, stabilizes the airplane in a straight-and-level attitude or, at the discretion of the examiner, reverses the direction of turn and repeats the maneuver in the opposite direction.
7. Avoids any indication of an approaching stall, abnormal flight attitude, or exceeding any structural or operating limitation during any part of the maneuver.

B. TASK: APPROACHES TO STALLS

REFERENCES: FAR Part 61; AC 61-21; FSB Report; Pilot's Operating Handbook, AFM.

THREE approaches to stall are required, as follows (unless otherwise specified by the FSB Report):

1. One in the takeoff configuration (except where the airplane uses only zero-flap takeoff configuration) or approach configuration.
2. One in a clean configuration.
3. One in a landing configuration.

One of these approaches to a stall must be accomplished while in a turn using a bank angle of 15 to 30°.

Objective. To determine that the applicant:

1. Exhibits adequate knowledge of the factors which influence stall characteristics, including the use of various drag configurations, power settings, pitch attitudes, weights, and bank angles. Also, exhibits adequate knowledge of the proper procedure for resuming normal flight.

2. Selects an entry altitude, when accomplished in an airplane, that is in accordance with the AFM or Operating Handbook, but in no case lower than an altitude that will allow recovery to be safely completed at a minimum of 3,000 feet (900 meters) AGL. When accomplished in an FTD or flight simulator, the entry altitude may be at low, intermediate, or high altitude as appropriate for the airplane and the configuration, at the discretion of the examiner.

3. Observes the area is clear of other aircraft prior to accomplishing an approach to a stall.

4. While maintaining altitude, slowly establishes the pitch attitude (using trim or elevator/stabilizer), bank angle, and power setting that will induce stall at the desired target airspeed.

5. Announces the first indication of an impending stall (such as buffeting, stick shaker, decay of control effectiveness, and any other cues related to the specific airplane design characteristics) and initiates recovery or as directed by the examiner (using maximum power or as directed by the examiner).

6. Recovers to a reference airspeed, altitude and heading, allowing only the acceptable altitude or airspeed loss, and heading deviation.

7. Demonstrates smooth, positive airplane control during entry, approach to a stall, and recovery.

C. TASK: POWERPLANT FAILURE — MULTIENGINE AIRPLANE

REFERENCES: FAR Part 61; Pilot's Operating Handbook, AFM.

NOTE: When not in an FTD or a flight simulator, the feathering of one propeller must be demonstrated in any multiengine airplane equipped with propellers which can be safely feathered and unfeathered while airborne. In a multiengine jet airplane, one engine must be shut down and a restart must be demonstrated while airborne. Feathering or shutdown should be performed only under conditions, and at such altitudes (no lower than 3,000 feet [900 meters] AGL) and in a position where a safe landing can be made on an established airport in the event difficulty is encountered in unfeathering the propeller or restarting the engine. At an altitude lower than 3,000 feet (900 meters) AGL, simulated engine failure will be performed by setting the powerplant controls to simulate zero-thrust. In the event propeller cannot be unfeathered or engine air started during the test, it should be treated as an emergency.

When authorized and conducted in a flight simulator, feathering or shutdown may be performed in conjunction with any procedure or maneuver and at locations and altitudes at the discretion of the examiner. However, when conducted in an FTD, authorizations shall be limited to shutdown, feathering, restart, and/or unfeathering procedures only. See Appendix 1.

Objective. To determine that the applicant:

1. Exhibits adequate knowledge of the flight characteristics and controllability associated with maneuvering with powerplant(s) inoperative (as appropriate to the airplane).
2. Maintains positive airplane control. Establishes a bank of approximately 5°, if required, or as recommended by the manufacturer, to maintain coordinated flight, and properly trims for that condition.
3. Sets powerplant controls, reduces drag as necessary, correctly identifies and verifies the inoperative powerplant(s) after the failure (or simulated failure).
4. Maintains the operating powerplant(s) within acceptable operating limits.
5. Follows the prescribed airplane checklist, and verifies the procedures for securing the inoperative powerplant(s).
6. Determines the cause for the powerplant(s) failure and if a restart is a viable option.

7. Maintains desired altitude within ±100 feet (30 meters), when a constant altitude is specified and is within the capability of the airplane.

8. Maintains the desired airspeed within ±10 knots.

9. Maintains the desired heading within ±10° of the specified heading.

10. Demonstrates proper powerplant restart procedures (if appropriate) in accordance with FAA approved procedure/checklist or the manufacturer's recommended procedures and pertinent checklist items.

D. TASK: POWERPLANT FAILURE — SINGLE–ENGINE AIRPLANE

REFERENCES: FAR Part 61; AC 61-21; Pilot's Operating Handbook, AFM.

NOTE: No simulated powerplant failure shall be given by the examiner in an airplane when an actual touchdown could not be safely completed should it become necessary.

Objective. To determine that the applicant:

1. Exhibits adequate knowledge of the flight characteristics, approach and forced (emergency) landing procedures, and related procedures to use in the event of a powerplant failure (as appropriate to the airplane).

2. Maintains positive airplane control throughout the maneuver.

3. Establishes and maintains the recommended best glide airspeed, ±5 knots, and configuration during a simulated powerplant failure.

4. Selects a suitable airport or landing area which is within the performance capability of the airplane.

5. Establishes a proper flight pattern to the selected airport or landing area, taking into account altitude, wind, terrain, obstructions, and other pertinent operational factors.

6. Follows the emergency checklist items appropriate to the airplane.

7. Determines the cause for the simulated powerplant failure (if altitude permits) and if a restart is a viable option.

8. Uses airplane configuration devices such as landing gear and flaps in a manner recommended by the manufacturer and/or approved by the FAA.

E. TASK: SPECIFIC FLIGHT CHARACTERISTICS

REFERENCES: FAR Part 61; FSB Report; Pilot's Operating Handbook, AFM.

Objective. To determine that the applicant:

1. Exhibits adequate knowledge of specific flight characteristics appropriate to the specific airplane, as identified by the FSB Report, such as Dutch Rolls in a Boeing 727 or Lear Jet.
2. Uses proper technique to enter into, operate within, and recover from specific flight situations.

V. INSTRUMENT PROCEDURES

NOTE: TASKS B through F are not required if the applicant holds a private pilot or commercial pilot certificate and is seeking a type rating limited to VFR. If TASK D, Nonprecision Instrument Approach Procedures, is performed in a training device (other than an FTD or flight simulator) and the applicant has completed an approved training course for the airplane type involved, not more than one (1) of the required instrument procedures may be observed by a person qualified to act as an instructor or check airman under that approved training program. The instrument approaches are considered to begin when the airplane is over the initial approach fix for the procedure being used and end when the airplane touches down on the runway or when transition to a missed approach configuration is completed. Instrument conditions need NOT be simulated below the minimum altitude for the approach being accomplished.

A. TASK: INSTRUMENT ARRIVAL

REFERENCES: FAR Part 61; Pilot's Operating Handbook, AFM, AIM; En Route Low/High Altitude Charts, Profile Descent Charts, STARs, Instrument Approach Procedure Charts.

Objective. To determine that the applicant:

1. Exhibits adequate knowledge of En Route Low and High Altitude Charts, STARs, Instrument Approach Charts, and related pilot and controller responsibilities.
2. Uses the current and appropriate navigation publications for the proposed flight.
3. Selects, and correctly identifies all instrument references, flight director and autopilot controls, and navigation and communications equipment associated with the arrival.
4. Performs the airplane checklist items appropriate to the arrival.
5. Establishes communications with ATC, using proper phraseology.
6. Complies, in a timely manner, with all ATC clearances, instructions, and restrictions.
7. Exhibits adequate knowledge of two-way communications failure procedures.
8. Intercepts, in a timely manner, all courses, radials, and bearings appropriate to the procedure, route, ATC clearance, or as directed by the examiner.
9. Adheres to airspeed restrictions and adjustments required by regulations, ATC, the pilot operating handbook, the AFM, or the examiner.

10. Establishes, where appropriate, a rate of descent consistent with the airplane operating characteristics and safety.
11. Maintains the appropriate airspeed/V-speed within ±10 knots, but not less than V_{REF}, if applicable; heading ±10°; altitude within ±100 feet (30 meters); and accurately tracks radials, courses, and bearings.
12. Complies with the provisions of the Profile Descent, STAR, and other arrival procedures, as appropriate.

B. TASK: HOLDING

REFERENCES: FAR Part 61; Pilot's Operating Handbook, AFM, AIM; En Route Low/High Altitude Charts, STARs, Instrument Approach Procedure Charts.

Objective. To determine that the applicant:

1. Exhibits adequate knowledge of holding procedures for standard and non-standard, published and non-published holding patterns. If appropriate, demonstrates adequate knowledge of holding endurance, including, but not necessarily limited to, fuel on board, fuel flow while holding, fuel required to alternate, etc.
2. Changes to the recommended holding airspeed appropriate for the airplane and holding altitude, so as to cross the holding fix at or below maximum holding airspeed.
3. Recognizes arrival at the clearance limit or holding fix.
4. Follows appropriate entry procedures for a standard, non-standard, published, or non-published holding pattern.
5. Complies with ATC reporting requirements.
6. Uses the proper timing criteria required by the holding altitude and ATC or examiner's instructions.
7. Complies with the holding pattern leg length when a DME distance is specified.
8. Uses the proper wind-drift correction techniques to maintain the desired radial, track, or bearing.
9. Arrives over the holding fix as close as possible to the "expect further clearance" time.
10. Maintains the appropriate airspeed/V-speed within ±10 knots, altitude within ±100 feet (30 meters), headings within ±10°; and accurately tracks radials, courses, and bearings.

C. TASK: PRECISION INSTRUMENT APPROACHES

REFERENCES: FAR Part 61; AC 61-27; Pilot's Operating Handbook, AFM, AIM; Instrument Approach Procedure Charts.

NOTE: Two precision approaches, utilizing airplane NAVAID equipment for centerline and glideslope guidance, must be accomplished in simulated instrument conditions to 200 feet above the runway/touchdown zone elevation. At least one approach must be flown manually. The second approach may be flown via the autopilot, if appropriate, and if the 200-foot altitude does not violate the authorized minimum altitude for autopilot operation. Manually flown precision approaches may use raw data displays or may be flight director assisted, at the discretion of the examiner.

For multiengine airplanes at least one manually controlled precision approach must be accomplished with a simulated failure of one powerplant. The simulated powerplant failure should occur before initiating the final approach segment and must continue to touchdown or throughout the missed approach procedure.

As the markings on localizer/glide slope indicators vary, a one-quarter scale deflection of either the localizer, or glide slope indicator is when it is displaced one-fourth of the distance that it may be deflected from the on glide slope or on localizer position.

Objective. To determine that the applicant:

1. Exhibits adequate knowledge of the precision instrument approach procedures with all engines operating, and with one engine inoperative.
2. Accomplishes the appropriate precision instrument approaches as selected by the examiner.
3. Establishes two-way communications with ATC using the proper communications phraseology and techniques, either personally, or, if appropriate, directs co-pilot/safety pilot to do so, as required for the phase of flight or approach segment.
4. Complies, in a timely manner, with all clearances, instructions, and procedures.
5. Advises ATC anytime the applicant is unable to comply with a clearance.
6. Establishes the appropriate airplane configuration and airspeed/V-speed considering turbulence, wind shear, microburst conditions, or other meteorological and operating conditions.

7. Completes the airplane checklist items appropriate to the phase of flight or approach segment, including engine out approach and landing checklists, if appropriate.

8. Prior to beginning the final approach segment, maintains the desired altitude ±100 feet (30 meters), the desired airspeed within ±10 knots, the desired heading within ±5°; and accurately tracks radials, courses, and bearings.

9. Selects, tunes, identifies, and monitors the operational status of ground and airplane navigation equipment used for the approach.

10. Applies the necessary adjustments to the published Decision Height and visibility criteria for the airplane approach category as required, such as —

    a. Notices to Airmen, including Flight Data Center Procedural NOTAMs.
    b. inoperative airplane and ground navigation equipment.
    c. inoperative visual aids associated with the landing environment.
    d. National Weather Service (NWS) reporting factors and criteria.

11. Establishes a predetermined rate of descent at the point where the electronic glide slope begins which approximates that required for the airplane to follow the glide slope.

12. Maintains a stabilized final approach, arriving at Decision Height with no more than one-quarter scale deflection of the localizer, or the glide slope indicators and the airspeed/V-speed within ±5 knots of that desired.

13. Avoids descent below the Decision Height before initiating a missed approach procedure or transitioning to a landing.

14. Initiates immediately the missed approach when at the Decision Height, and the required visual references for the runway are not distinctly visible and identifiable.

15. Transitions to a normal landing approach only when the airplane is in a position from which a descent to a landing on the runway can be made at a normal rate of descent using normal maneuvering.

16. Maintains localizer and glide slope within one-quarter scale deflection of the indicators during the visual descent from Decision Height to a point over the runway where glide slope must be abandoned to accomplish a normal landing.

D. TASK: NONPRECISION INSTRUMENT APPROACHES

REFERENCES: FAR Part 61; AC 61-27; Pilot's Operating Handbook, AFM, AIM; Instrument Approach Procedure Charts.

NOTE: The applicant must accomplish at least two nonprecision approaches. The examiner will select nonprecision approaches that are representative of that which the applicant is likely to use. The second nonprecision approach will utilize a navigational aid other than the one used for the first approach.

Objective. To determine that the applicant:

1. Exhibits adequate knowledge of nonprecision approach procedures representative of those the applicant is likely to use.
2. Accomplishes the nonprecision instrument approaches selected by the examiner.
3. Establishes two-way communications with ATC as appropriate to the phase of flight or approach segment and uses proper communications phraseology and techniques.
4. Complies with all clearances issued by ATC.
5. Advises ATC or the examiner any time the applicant is unable to comply with a clearance.
6. Establishes the appropriate airplane configuration and airspeed, and completes all applicable checklist items.
7. Maintains, prior to beginning the final approach segment, the desired altitude ±100 feet (30 meters), the desired airspeed ±10 knots, the desired heading ±5°; and accurately tracks radials, courses, and bearings.
8. Selects, tunes, identifies, and monitors the operational status of ground and airplane navigation equipment used for the approach.
9. Applies the necessary adjustments to the published Minimum Descent Altitude and visibility criteria for the airplane approach category when required, such as —

    a. Notices to Airmen, including Flight Data Center Procedural NOTAMs.
    b. inoperative airplane and ground navigation equipment.
    c. inoperative visual aids associated with the landing environment.
    d. National Weather Service (NWS) reporting factors and criteria.

10.   Establishes a rate of descent that will ensure arrival at the Minimum Descent Altitude (at, or prior to reaching, the visual descent point (VDP), if published) with the airplane in a position from which a descent from MDA to a landing on the intended runway can be made at a normal rate using normal maneuvering.

11.   Allows, while on the final approach segment, not more than quarter-scale deflection of the Course Deviation Indicator (CDI) or ±5° in the case of the RMI or bearing pointer, and maintains airspeed within ±5 knots of that desired.

12.   Maintains the Minimum Descent Altitude, when reached, within −0, +50 feet (−0, +15 meters) to the missed approach point.

13.   Executes the missed approach if the required visual references for the intended runway are not distinctly visible and identifiable at the missed approach point.

14.   Executes a normal landing from a straight-in or circling approach when instructed by the examiner.

E. TASK: CIRCLING APPROACH

REFERENCES: FAR Part 61; AC 61-27; Pilot's Operating Handbook, AFM, AIM; Instrument Approach Procedure Charts.

Objective. To determine that the applicant:

1.   Exhibits adequate knowledge of circling approach categories, speeds, and procedures to a specified runway.

2.   Accomplishes the circling approach selected by the examiner.

3.   Demonstrates sound judgment and knowledge of the airplane maneuvering capabilities throughout the circling approach.

4.   Confirms the direction of traffic and adheres to all restrictions and instructions issued by ATC.

5.   Descends at a rate that ensures arrival at the MDA at, or prior to, a point from which a normal circle-to-land maneuver can be accomplished.

6.   Avoids descent below the appropriate circling Minimum Descent Altitude or exceeding the visibility criteria until in a position from which a descent to a normal landing can be made.

7.   Maneuvers the airplane, after reaching the authorized circling approach altitude, by visual references to maintain a flightpath that permits a normal landing on a runway at least 90° from the final approach course.

8.   Performs the procedure without excessive maneuvering and without exceeding the normal operating limits of the airplane (the angle of bank should not exceed 30°).

9.   Maintains the desired altitude within –0, +100 feet (–0, +30 meters), heading/track within ±5°, the airspeed/V-speed within ±5 knots, but not less than the airspeed as specified in the pilot operating handbook or the AFM.

10. Uses the appropriate airplane configuration for normal and abnormal situations and procedures.

11. Turns in the appropriate direction, when a missed approach is dictated during the circling approach, and uses the correct procedure and airplane configuration.

12. Performs all procedures required for the circling approach and airplane control in a smooth, positive, and timely manner.

F. TASK: MISSED APPROACH

REFERENCES: FAR Part 61; AC 61-27; Pilot's Operating Handbook, AFM, AIM; Instrument Approach Procedure Charts.

NOTE: The applicant must be required to perform at least two missed approaches with at least one missed approach from a precision approach (ILS, MLS, or GPS). A complete approved missed approach must be accomplished at least once. Additionally, in multiengine airplanes, a missed approach must be accomplished with one engine inoperative (or simulated inoperative). The engine failure may be experienced anytime prior to the initiation of the approach, during the approach, or during the transition to the missed approach attitude and configuration.

Going below the MDA or DH, as appropriate, prior to the initiation of the missed approach shall be considered unsatisfactory performance. However, satisfactory performance may be concluded if the missed approach is properly initiated at DH and the airplane descends below DH only because of the momentum of the airplane transitioning from a stabilized approach to a missed approach.

Objective. To determine that the applicant:

1.   Exhibits adequate knowledge of missed approach procedures associated with standard instrument approaches.

2.   Initiates the missed approach procedure promptly by the timely application of power, establishes the proper climb attitude, and reduces drag in accordance with the approved procedures.

3.   Reports to ATC, beginning the missed approach procedure.

4. Complies with the appropriate missed approach procedure or ATC clearance.

5. Advises ATC any time the applicant is unable to maneuver the airplane to comply with a clearance.

6. Follows the recommended airplane checklist items appropriate to the go-around procedure for the airplane used.

7. Requests clearance, if appropriate, to the alternate airport, another approach, a holding fix, or as directed by the examiner.

8. Maintains the desired altitudes ±100 feet (30 meters), airspeed ±5 knots, heading ±5°; and accurately tracks courses, radials, and bearings.

VI. LANDINGS AND APPROACHES TO LANDINGS

NOTE: Notwithstanding the authorizations for the combining of maneuvers and for the waiver of maneuvers, the applicant must make at least three (3) actual landings (one to a full stop). These landings must include the types listed in this AREA OF OPERATION; however, more than one type may be combined where appropriate (i.e., crosswind and landing from a precision approach or landing with simulated powerplant failure, etc.).

For all landings in airplanes, touchdown should be 500 to 3,000 feet (150 to 900 meters) past the runway threshold, not to exceed one-third of the runway length, with the runway centerline between the main gear.

A. TASK: NORMAL AND CROSSWIND APPROACHES AND LANDINGS

REFERENCES: FAR Part 61; AC 61-21; Pilot's Operating Handbook, AFM.

NOTE: In an airplane with a single powerplant, unless the applicant holds a commercial pilot certificate, he/she must accomplish accuracy approaches and spot landings from an altitude of 1,000 feet (300 meters) or less, with the engine power lever in idle and 180° of change in direction. The airplane must touch the ground in a normal landing attitude beyond and within 200 feet (60 meters) of a designated line or point on the runway. At least one landing must be from a forward slip. Although circular approaches are acceptable, 180° approaches using two 90° turns with a straight base leg are preferred.

Objective. To determine that the applicant:

1. Exhibits adequate knowledge of normal and crosswind approaches and landings including recommended approach angles, airspeeds, V-speeds, configurations, performance limitations, wake turbulence, and safety factors (as appropriate to the airplane).
2. Establishes the approach and landing configuration appropriate for the runway and meteorological conditions, and adjusts the powerplant controls as required.
3. Maintains a ground track that ensures the desired traffic pattern will be flown, taking into account any obstructions and ATC or examiner instructions.
4. Verifies existing wind conditions, makes proper correction for drift, and maintains a precise ground track.

5. Maintains a stabilized approach and the desired airspeed/V-speed within ±5 knots.
6. Accomplishes a smooth, positively controlled transition from final approach to touchdown.
7. Maintains positive directional control and crosswind correction during the after-landing roll.
8. Uses spoilers, prop reverse, thrust reverse, wheel brakes, and other drag/braking devices, as appropriate, in such a manner to bring the airplane to a safe stop.
9. Completes the applicable after-landing checklist items in a timely manner and as recommended by the manufacturer.

B. TASK: LANDING FROM A PRECISION APPROACH

REFERENCES: FAR Part 61; AC 61-27; Pilot's Operating Handbook, AFM, AIM.

NOTE: If circumstances beyond the control of the applicant prevent an actual landing, the examiner may accept an approach to a point where, in his/her judgment, a safe landing and a full stop could have been made. Where a simulator, approved for landing from a precision approach, is used, the approach may be continued through the landing and credit given for one of the landings required by this AREA OF OPERATION.

Objective. To determine that the applicant:

1. Exhibits awareness of landing in sequence from a precision approach.
2. Considers factors to be applied to the approach and landing such as displaced thresholds, meteorological conditions, NOTAMs, and ATC or examiner instructions.
3. Uses the airplane configuration and airspeed/V-speeds, as appropriate.
4. Maintains, during the final approach segment, glide slope and localizer indications within applicable standards of deviation, and the recommended airspeed/V-speed ±5 knots.
5. Applies gust/wind factors as recommended by the manufacturer, and takes into account meteorological phenomena such as wind shear, microburst, and other related safety of flight factors.
6. Accomplishes the appropriate checklist items.
7. Transition smoothly from simulated instrument meteorological conditions at a point designated by the examiner, maintaining positive airplane control.

8. Accomplishes a smooth, positively controlled transition from final approach to touchdown.

9. Maintains positive directional control and crosswind correction during the after-landing roll.

10. Uses spoilers, prop reverse, thrust reverse, wheel brakes, and other drag/braking devices, as appropriate, in such a manner to bring the airplane to a safe stop after landing.

11. Completes the after-landing checklist items in a timely manner and as recommended by the manufacturer.

C. TASK: APPROACH AND LANDING WITH (SIMULATED) POWERPLANT FAILURE — MULTIENGINE AIRPLANE

REFERENCES: FAR Part 61; AC 61-21; Pilot's Operating Handbook, AFM.

NOTE: In airplanes with three powerplants, the applicant shall follow a procedure (if approved) that approximates the loss of two powerplants, the center and one outboard powerplant. In other multiengine airplanes, the applicant shall follow a procedure which simulates the loss of 50 percent of available powerplants, the loss being simulated on one side of the airplane.

Objective. To determine that the applicant:

1. Exhibits adequate knowledge of the flight characteristics and controllability associated with maneuvering to a landing with (a) powerplant(s) inoperative (or simulated inoperative) including the controllability factors associated with maneuvering, and the applicable emergency procedures.

2. Maintains positive airplane control. Establishes a bank of approximately 5°, if required, or as recommended by the manufacturer, to maintain coordinated flight, and properly trims for that condition.

3. Sets powerplant controls, reduces drag as necessary, correctly identifies and verifies the inoperative powerplant(s) after the failure (or simulated failure).

4. Maintains the operating powerplant(s) within acceptable operating limits.

5. Follows the prescribed airplane checklist, and verifies the procedures for securing the inoperative powerplant(s).

6. Proceeds toward the nearest suitable airport.

7. Maintains, prior to beginning the final approach segment, the desired altitude ±100 feet (30 meters), the desired airspeed ±10 knots, the desired heading ±5°; and accurately tracks courses, radials, and bearings.

8. Establishes the approach and landing configuration appropriate for the runway or landing area, and meteorological conditions; and adjusts the powerplant controls as required.
9. Maintains a stabilized approach and the desired airspeed/V-speed within ±5 knots.
10. Accomplishes a smooth, positively-controlled transition from final approach to touchdown.
11. Maintains positive directional control and crosswind corrections during the after-landing roll.
12. Uses spoilers, prop reverse, thrust reversers, wheel brakes and other drag/braking devices, as appropriate, in such a manner to bring the airplane to a safe stop after landing
13. Completes the after-landing checklist items in a timely manner, after clearing the runway, and as recommended by the manufacturer.

D. TASK: LANDING FROM A CIRCLING APPROACH

REFERENCES: FAR Part 61; AC 61-27; Pilot's Operating Handbook, AFM, AIM.

Objective. To determine that the applicant:

1. Exhibits adequate knowledge of a landing from a circling approach.
2. Selects, and complies with, a circling approach procedure to a specified runway.
3. Considers the environmental, operational, and meteorological factors which affect a landing from a circling approach.
4. Confirms the direction of traffic and adheres to all restrictions and instructions issued by ATC.
5. Descends at a rate that ensures arrival at the MDA at, or prior to, a point from which a normal circle-to-land maneuver can be accomplished.
6. Avoids descent below the appropriate circling MDA or exceeding the visibility criteria until in a position from which a descent to a normal landing can be made.
7. Accomplishes the appropriate checklist items.
8. Maneuvers the airplane, after reaching the authorized circling approach altitude, by visual references, to maintain a flightpath that permits a normal landing on a runway at least 90° from the final approach course.
9. Performs the maneuver without excessive maneuvering and without exceeding the normal operating limits of the airplane. The angle of bank should not exceed 30°.

10. Maintains the desired altitude within +100, −0 feet (+30, −0 meters), heading within ±5°, and approach airspeed/V-speed within ±5.
11. Uses the appropriate airplane configuration for normal and abnormal situations and procedures.
12. Performs all procedures required for the circling approach and airplane control in a timely, smooth, and positive manner.
13. Accomplishes a smooth, positively controlled transition to final approach and touchdown.
14. Maintains positive directional control and crosswind correction during the after-landing roll.
15. Uses spoilers, prop reverse, thrust reverse, wheel brakes, and other drag/braking devices, as appropriate, in such a manner to bring the airplane to a safe stop.
16. Completes the after-landing checklist items, after clearing the runway, in a timely manner and as recommended by the manufacturer.

E. TASK: REJECTED LANDING

REFERENCES: FAR Part 61; AC 61-21; Pilot's Operating Handbook, AFM; FSB Report.

NOTE: The maneuver may be combined with instrument, circling, or missed approach procedures, but instrument conditions need not be simulated below 100 feet (30 meters) above the runway. This maneuver should be initiated approximately 50 feet (15 meters) above the runway and approximately over the runway threshold or as recommended by the FSB Report.

For those applicants seeking a VFR only type rating in an airplane not capable of instrument flight, for those cases where this maneuver is accomplished with a simulated engine failure, it should not be initiated at speeds or altitudes below that recommended in the pilot's operating manual.

Objective. To determine that the applicant:

1. Exhibits adequate knowledge of a rejected landing procedure, including the conditions that dictate a rejected landing, the importance of a timely decision, the recommended airspeed/V-speeds, and also the applicable "clean-up" procedure.
2. Makes a timely decision to reject the landing for actual or simulated circumstances and makes appropriate notification when safety-of-flight is not an issue.

3. Applies the appropriate power setting for the flight condition and establishes a pitch attitude necessary to obtain the desired performance.
4. Retracts the wing flaps/drag devices and landing gear, if appropriate, in the correct sequence and at a safe altitude, establishes a positive rate of climb and the appropriate airspeed/V-speed within ±5 knots.
5. Trims the airplane as necessary, and maintains the proper ground track during the rejected landing procedure.
6. Accomplishes the appropriate checklist items in a timely manner in accordance with approved procedures.

F. TASK: LANDING FROM A ZERO OR NONSTANDARD FLAP APPROACH

REFERENCES: FAR Part 61; AC 61-21; FSB Report; Pilot's Operating Handbook, AFM.

NOTE: This maneuver need not be accomplished for a particular airplane type if the Administrator has determined that the probability of flap extension failure on that type airplane is extremely remote due to system design. The examiner must determine whether checking on slats only and partial-flap approaches are necessary for the practical test.

Objective. To determine that the applicant:

1. Exhibits adequate knowledge of the factors which affect the flight characteristics of an airplane when full or partial flaps, leading edge flaps, and other similar devices become inoperative.
2. Uses the correct airspeeds/V-speeds for the approach and landing.
3. Maintains the proper airplane pitch attitude and flightpath for the configuration, gross weight, surface winds, and other applicable operational considerations.
4. Uses runway of sufficient length for the zero or nonstandard flap condition.
5. Maneuvers the airplane to a point where, in the opinion of the examiner, touchdown at an acceptable point on the runway and a safe landing to a full stop could be made.
6. If a landing is made, uses spoilers, prop reverse, thrust reverse, wheel brakes, and other drag/braking devices, as appropriate, in such a manner to bring the airplane to a safe stop.

VII.   NORMAL AND ABNORMAL PROCEDURES

REFERENCES: FAR Part 61; Pilot's Operating Handbook, AFM.

Objective. To determine that the applicant:

1.   Possesses adequate knowledge of the normal and abnormal procedures of the systems, subsystems, and devices relative to the airplane type (as may be determined by the examiner) knows immediate action items to accomplish, if appropriate, and proper checklist to accomplish or to call for, if appropriate.
2.   Demonstrates the proper use of the airplane systems, subsystems, and devices (as may be determined by the examiner) appropriate to the airplane such as —

   a.   powerplant.
   b.   fuel system.
   c.   electrical system.
   d.   hydraulic system.
   e.   environmental and pressurization systems.
   f.   fire detection and extinguishing systems.
   g.   navigation and avionics systems.
   h.   automatic flight control system, electronic flight instrument system, and related subsystems.
   i.   flight control systems.
   j.   anti-ice and deice systems.
   k.   airplane and personal emergency equipment.
   l.   other systems, subsystems, and devices specific to the type airplane, including make, model, and series.

VIII. EMERGENCY PROCEDURES

REFERENCES: FAR Part 61; Pilot's Operating Handbook, AFM.

Objective. To determine that the applicant:

1. Possesses adequate knowledge of the emergency procedures (as may be determined by the examiner) relating to the particular airplane type.

2. Demonstrates the proper emergency procedures (as must be determined by the examiner) relating to the particular airplane type, including —

   a. emergency descent (maximum rate).
   b. inflight fire and smoke removal.
   c. rapid decompression.
   d. emergency evacuation.
   e. others (as may be required by the AFM)

3. Demonstrates the proper procedure for any other emergency outlined (as must be determined by the examiner) in the appropriate approved AFM.

IX. POSTFLIGHT PROCEDURES

A. TASK: AFTER LANDING

REFERENCES: Pilot's Operating Handbook, AFM.

Objective. To determine that the applicant:

1. Exhibits adequate knowledge of safe after-landing/taxi procedures as appropriate.
2. Demonstrates proficiency by maintaining correct and positive control. In airplanes equipped with float devices, this includes water taxiing, approaching a buoy, and docking.
3. Maintains proper spacing on other aircraft, obstructions, and persons.
4. Accomplishes the applicable checklist items and performs the recommended procedures.
5. Maintains the desired track and speed.
6. Complies with instructions issued by ATC (or the examiner simulating ATC).
7. Observes runway hold lines, localizer and glide slope critical areas, and other surface control markings and lighting.
8. Maintains constant vigilance and airplane control during the taxi operation.

B. TASK: PARKING AND SECURING

REFERENCES: Pilot's Operating Handbook, AFM.

Objective. To determine that the applicant:

1. Exhibits adequate knowledge of the parking and the securing airplane procedures.
2. Applicant has adequate knowledge of the airplane forms/logs to record the flight time/discrepancies.

Appendix 1 - Airplanes

TASK VS. SIMULATION DEVICE CREDIT

Examiners conducting the Airline Transport Pilot and/or Type Rating Practical Tests with simulation devices should consult appropriate documentation to ensure that the device has been approved for training and checking the TASKS in question. The documentation for each device should reflect that the following activities have occurred:

    1. The device must be evaluated, determined to meet the appropriate standards, and assigned the appropriate qualification level by the National Simulator Program Manager. The device must continue to meet qualification standards through continuing evaluations as outlined in the appropriate advisory circular (AC). For airplane flight training devices (FTD's), AC 120-45(as amended), Airplane Flight Training Device Qualification, will be used. For simulators, AC 120-40 (as amended), Airplane Simulator Qualification, will be used.

    2. The FAA must approve the device for specific TASKS.

    3. The device must continue to support the level of student or applicant performance required by this PTS.

NOTE: Users of the following chart are cautioned that use of the chart alone is incomplete. The description and objective of each task as listed in the body of the PTS, including all notes, must also be incorporated for accurate simulation device use.

USE OF CHART

X Creditable.

A Creditable if appropriate systems are installed and operating.

NOTE: 1. The airplane may be used for all tasks.
2. Level C simulators may be used as indicated only if the applicant meets established pre-requisite experience requirements.
3. Training Devices below Level 4 may NOT be used for Airplane Type Ratings.
4. Standards for and use of Level 1 Flight Training Devices have not been determined

FLIGHT TASK

Areas of Operation : Section Two

	FLIGHT SIMULATION DEVICE LEVEL										
	1	2	3	4	5	6	7	A	B	C	D
I. Preflight Procedures											
A. Preflight Inspection (Cockpit Only)		A	X					X	X	X	X
II. Ground Operations											
A. Powerplant Start		A	X	A	A	A	X	X	X	X	X
B. Taxiing										X	X
C. Pretakeoff Checks		A	X	A	A	A	X	X	X	X	X
III. Takeoff and Departure Maneuvers											
A. Normal and Crosswind Takeoff								X	X	X	X
B. Instrument Takeoff (Levels 3, 6, &7 require a visual sys. approved in accordance with AC 120-40, as amended)			X			X	X	X	X	X	X
C. Powerplant Failure During Takeoff								X	X	X	X
D. Rejected Takeoff (Levels 3, 6, & 7 require a visual sys. approved in accordance with AC 120-40, as amended)			X			X	X	X	X	X	X
E. Instrument Departure			X			X	X	X	X	X	X
IV. Inflight Maneuvers											
A. Steep Turns						X	X	X	X	X	X
B. Approaches to Stalls (Use of Levels 3, 6, & 7 require operational synthetic stall warning system)			X			X	X	X	X	X	X
C. Powerplant Failure - Multiengine Airplane							X	X	X	X	X
D. Powerplant Failure - Single-Engine Airplane				X			X	X	X	X	X
E. Specific Flight Characteristics	Level of device as determined by the airplane Flight Standardization Board (FSB)										

FLIGHT TASK Areas of Operation : Section Two	FLIGHT SIMULATION DEVICE LEVEL										
	1	2	3	4	5	6	7	A	B	C	D
V. Instrument Procedures											
A. Instrument Arrival			X			X	X	X	X	X	X
B. Holding			X			X	X	X	X	X	X
C1. Precision Approach (All Eng. Operating) (Autopilot/Manual Flt. Dir. Assist/Manual Raw Data) (Levels 2&5 use limited to A/P coupled approach only)			A	X		A	A	X	X	X	X
C2. Precision Approach (One Eng. Inop.) (Manual Flt. Dir. Asst/Manual Raw Data)								X	X	X	X
D. Nonprecision Approach (Not more than one authorized in a device less than Level A simulator) (Levels 2&5 use limited to A/P coupled approach only)			A	X		A	A	X	X	X	X
E. Circling Approach (each appr. must be specifically auth.)									X	X	X
F1. Missed Approach (Normal)			X			X	X	X	X	X	X
F2. Missed Approach (Powerplant Failure)							X	X	X	X	X
VI. Landings and Approaches to Landings											
A. Normal and Crosswind Approaches and Landings										X	X
B. Landing From a Precision Approach										X	X
C. Landing with Powerplant Failure										X	X
D. Landing From Circling Approach										X	X
E. Rejected Landing								X	X	X	X
F. Landing From 0° or Nonstandard Flap Approach										X	X

Areas of Operation : Section Two	FLIGHT SIMULATION DEVICE LEVEL										
	1	2	3	4	5	6	7	A	B	C	D
VII. Normal and Abnormal Procedures (*1) (*2)											
A. Powerplant (including shutdown & restart)		A	X	A	A	X	X	X	X	X	X
B. Fuel System		A	X	A	A	X	X	X	X	X	X
C. Electrical System		A	X	A	A	X	X	X	X	X	X
D. Hydraulic System		A	X	A	A	X	X	X	X	X	X
E. Environmental and Pressurization Systems		A	X	A	A	X	X	X	X	X	X
F. Fire Detection and Extinguisher Systems		A	X	A	A	X	X	X	X	X	X
G. Navigation and Avionics Systems		A	X	A	A	X	X	X	X	X	X
H. Automatic Flight Control System, Electronic Flight Instrument System, and Related Subsystem		A	X	A	A	X	X	X	X	X	X
I. Flight Control Systems								X	X	X	X
J. Anti-ice and Deice Systems		A	X	A	A	A	X	X	X	X	X
K. Aircraft and Personal Emergency Equipment			A	X	A	A	X	X	X	X	X
L. Others, as determined by make, model, or series				A	A	X	X	X	X	X	X
VIII. Emergency Procedures											
A. Emergency Descent (Max. Rate)						X	X	X	X	X	X
B. Inflight Fire and Smoke Removal		A	X	A	A	X	X	X	X	X	X
C. Rapid Decompression		A	X	A	A	X	X	X	X	X	X
D. Emergency Evacuation				X				X	X	X	X
E. Others (as may be required by AFM)			A	X	A	A	X	X	X	X	X
IX. Postflight Procedures											
A. After Landing		A	X	A	A	X	X	X	X	X	X
B. Parking and Securing		A	X	A	A	X	X	X	X	X	X

(*1) Evaluation of normal and abnormal procedures may be accomplished in conjunction with other events.
(*2) Situations resulting in asymmetrical thrust or drag conditions (i.e., asymmetrical flight controls) must be accomplished in at least a Level A device. However, shutdown and restart (procedures only) may be accomplished in a properly equipped FTD.

Part 2 - Helicopters

CONTENTS

                FAA-S-8081-5B

PART 2, HELICOPTERS
AREA OF OPERATION: SECTION ONE

I. EQUIPMENT KNOWLEDGE

A. TASK: EQUIPMENT EXAMINATION

REFERENCES: FAR Part 61; Pilot's Operating Handbook, FAA Approved Rotorcraft Flight Manual (RFM).

Objective. To determine that the applicant:

1. Exhibits adequate knowledge appropriate to the helicopter; its systems and components; its normal, abnormal, and emergency procedures; and uses the correct terminology with regard to the following items —

    a. landing gear — indicators, brakes, tires, nosewheel steering, skids, and shocks.

    b. Powerplant — controls and indications, induction system, carburetor and fuel injection, exhaust and turbocharging, cooling, fire detection/protection, mounting points, turbine wheels, compressors, and other related components.

    c. fuel system — capacity; drains; pumps; controls; indicators; crossfeeding; transferring; jettison; fuel grade, color and additives; fueling and defueling procedures; and emergency substitutions, if applicable.

    d. oil system — capacity, grade, quantities, and indicators.

    e. hydraulic system — capacity, pumps, pressure, reservoirs, grade, and regulators.

    f. electrical system — alternators, generators, battery, circuit breakers and protection devices, controls, indicators, and external and auxiliary power sources and ratings.

    g. environmental systems — heating, cooling, ventilation, oxygen and pressurization, controls, indicators, and regulating devices.

    h. avionics and communications — autopilot; flight director; Electronic Flight Indicating Systems (EFIS); Flight Management System(s) (FMS); Long Range Navigation (LORAN) systems; Doppler Radar; Inertial Navigation Systems (INS); Global Positioning System (GPS/DGPS-/WGPS); VOR, NDB, ILS/MLS, RNAV systems and components; indicating devices; transponder; and emergency locator transmitter.

  i. ice protection — anti-ice, deice, pitot-static system protection, windshield, airfoil surfaces, and rotor protection.

  j. crewmember and passenger equipment — oxygen system, survival gear, emergency exits, evacuation procedures and crew duties, and quick donning oxygen mask for crewmembers and passengers.

  k. main/tail rotor systems — transmissions, gear boxes, oil/fluid levels, tolerances, and limitations.

2. Exhibits adequate knowledge of the contents of the Pilot Operating Handbook or RFM with regard to the systems and components listed in 1.; the Minimum Equipment List (MEL), if appropriate, and the Operations Specifications, if applicable.

B. TASK: PERFORMANCE AND LIMITATIONS

REFERENCES: FAR Parts 1, 61, 91; Pilot's Operating Handbook, RFM.

Objective. To determine that the applicant:

1. Exhibits adequate knowledge of performance and limitations, including a thorough knowledge of the adverse effects of exceeding any limitation.

2. Demonstrates proficient use of (as appropriate to the helicopter) performance charts, tables, graphs, or other data relating to items such as —

  a. takeoff performance: all engines, engine(s) inoperative.

  b. climb performance, all engines, engine(s) inoperative, and other engine malfunctions.

  c. service ceiling, all engines, engines(s) inoperative.

  d. cruise performance.

  e. fuel consumption, range, and endurance.

  f. descent performance.

  g. go-around from rejected landings.

  h. hovering in and out of ground effect.

  i. other performance data (appropriate to the helicopter).

3. Describes (as appropriate to the helicopter) the performance airspeeds used during specific phases of flight.

4. Describes the effects of meteorological conditions upon performance characteristics and correctly applies these factors to a specific chart, table, graph or other perform-ance data.
5. Computes the center-of-gravity location for a specific load condition (as specified by the examiner), including adding, removing, or shifting weight.
6. Determines if the computed center of gravity is within the forward, aft, and lateral (if applicable) center-of-gravity lim-its for takeoff and landing.
7. Demonstrates good planning and knowledge of proce-dures in applying operational factors affecting helicopter performance.

PART 2, HELICOPTERS
SECTION TWO: AREAS OF OPERATION

I. PREFLIGHT PROCEDURES

A. TASK: PREFLIGHT INSPECTION

REFERENCES: FAR Parts 61, 91; Pilot's Operating Handbook, RFM.

Objective. To determine that the applicant:

1. Exhibits adequate knowledge of the preflight inspection procedures, while explaining briefly —

   a. the purpose of inspecting the items which must be checked.
   b. how to detect possible defects.
   c. the corrective action to take.

2. Exhibits adequate knowledge of the operational status of the helicopter by locating and explaining the significance and importance of related helicopter documents such as —

   a. airworthiness and registration certificates.
   b. operating limitations, handbooks, and manuals.
   c. minimum equipment list (MEL) (if appropriate).
   d. weight and balance data.
   e. maintenance requirements, tests, and appropriate records applicable to the proposed flight or operation; and maintenance that may be performed by the pilot or other designated crewmember.

3. Uses the approved checklist to systematically inspect the helicopter externally and internally.
4. Uses the challenge-and-response (or other approved) method with the other crewmember(s), where applicable, to accomplish the checklist procedures.
5. Verifies the helicopter is safe for flight by emphasizing (as appropriate to the helicopter) the need to look at and explain the purpose of inspecting items such as —

   a. powerplant, including controls and indicators.
   b. fuel quantity, grade, type, contamination safeguards, and servicing procedures.
   c. oil quantity, grade, and type.

      d.   hydraulic fluid quantity, grade, type, and servicing procedures.

      e.   oxygen quantity, pressures, servicing procedures, and associated systems and equipment for crew and passengers.

      f.   skidtubes or landing gear, brakes, and steering system, where applicable.

      g.   tires for condition, inflation, and correct mounting, where applicable.

      h.   fire protection/detection systems for proper operation, servicing, pressures, and discharge indications.

      i.   pneumatic system pressures and servicing.

      j.   ground environmental systems for proper servicing and operation.

      k.   auxiliary power unit (APU) for servicing and operation.

      l.   flight control systems including trim, rotor blades, and associated components.

      m.   main rotor and anti-torque systems.

      n.   anti-ice, deice systems, servicing, and operation.

6. Coordinates with ground crew and ensures adequate clearance prior to moving any devices such as doors or hatches.

7. Complies with the provisions of the appropriate Operations Specifications, if applicable, as they pertain to the particular helicopter and operation.

8. Demonstrates proper operation and verification of all helicopter systems.

9. Notes any discrepancies, determines if the helicopter is airworthy and safe for flight, or takes the proper corrective action.

10. Checks the general area around the helicopter for hazards to the safety of the helicopter and personnel.

II. GROUND OPERATIONS

A. TASK: POWERPLANT START

REFERENCES: FAR Part 61; Pilot's Operating Handbook, RFM.

Objective. To determine that the applicant:

1. Exhibits adequate knowledge of the correct powerplant start procedures including the use of an external power source, starting under various atmospheric conditions, normal and abnormal starting limitations, and the proper action required in the event of a malfunction.
2. Ensures the ground safety procedures are followed during the before-start, start, and after-start phases.
3. Ensures the use of appropriate ground crew personnel during the start procedures.
4. Performs all items of the start procedures by systematically following the approved checklist items for the before-start, start, and after-start phases.
5. Demonstrates sound judgment and operating practices in those instances where specific instructions or checklist items are not published.

B. TASK: TAXIING

REFERENCES: FAR Part 61; Pilot's Operating Handbook, RFM.

Objective. To determine that the applicant:

1. Exhibits adequate knowledge of safe and appropriate taxi procedures.
2. Demonstrates proficiency by maintaining correct and positive helicopter control such as hover height, turns, and speed. This includes hovering taxi, air taxiing; and in helicopters with wheels, includes ground taxiing. In helicopters equipped with float devices, this includes water taxiing, approaching a buoy, and docking.
3. Maintains proper spacing on other aircraft and persons taking into consideration rotorwash and flying debris.
4. Accomplishes the applicable checklist items and performs recommended procedures.
5. Maintains desired and appropriate track and speed.
6. Complies with instructions issued by ATC (or the examiner simulating ATC).

7. Observes runway hold lines, localizer and glide slope critical areas, and other surface control markings and lighting.
8. Maintains constant vigilance and control of the helicopter during taxi operation.

C. TASK: PRETAKEOFF CHECKS

REFERENCES: FAR Part 61; Pilot's Operating Handbook, RFM.

Objective. To determine that the applicant:

1. Exhibits adequate knowledge of the pretakeoff checks by stating the reason for checking the items outlined on the approved checklist and explaining how to detect possible malfunctions.
2. Divides attention inside and outside cockpit.
3. Ensures that all systems are within their normal operating range prior to beginning, during the performance of, and at the completion of those checks required by the approved checklist.
4. Able to explain any normal or abnormal system operating characteristic or limitation; and the corrective action for a specific malfunction.
5. Determines if the helicopter is safe for the proposed flight or requires maintenance.
6. Determines the helicopter's takeoff performance, considering such factors as wind, density altitude, helicopter weight, temperature, pressure altitude, and departure route or routing.
7. Determines airspeeds/V-speeds and properly sets all instrument references, flight director and autopilot controls, and navigation and communications equipment.
8. Reviews procedures for emergency and abnormal situations which may be encountered during takeoff, and states the corrective action required of the pilot in command and other concerned crewmembers.
9. Obtains and correctly interprets the takeoff and departure clearance as issued by ATC.

III. TAKEOFF AND DEPARTURE MANEUVERS

A. TASK: NORMAL AND CROSSWIND TAKEOFF

REFERENCES: FAR Part 61; Pilot's Operating Handbook, RFM.

Objective. To determine that the applicant:

1. Exhibits adequate knowledge of normal and crosswind takeoffs and climbs including (as appropriate to the helicopter) airspeeds, configurations, and emergency/abnormal procedures. Performs all required pretakeoff checks as required by the appropriate checklist items.
2. Adjusts the powerplant controls as recommended by the FAA-approved guidance for the existing conditions.
3. Notes any obstructions or other hazards in the takeoff path.
4. Verifies and correctly applies the existing wind component to the takeoff performance.
5. Completes required checks prior to starting takeoff to verify the expected powerplant performance.
6. Aligns the helicopter on the runway centerline, or with the takeoff path.
7. Applies the controls correctly to maintain longitudinal alignment on the centerline of the runway or intended flightpath, prior to initiating and during the takeoff.
8. Increases power smoothly and positively to a predetermined value.
9. Monitors powerplant controls, settings, and instruments during takeoff to ensure all predetermined parameters are met.
10. Accelerates through effective translational lift to normal climb speed.
11. Uses the applicable noise abatement, wake turbulence avoidance procedures, as required.
12. Accomplishes the appropriate checklist items.
13. Maintains the appropriate climb segment airspeed/V-speeds.
14. Maintains the desired heading within ±5° and the desired airspeed/V-speed within ±5 knots.

B. TASK: INSTRUMENT TAKEOFF

REFERENCES: FAR Part 61; AC 61-27; Pilot's Operating Handbook, RFM, AIM.

Objective. To determine that the applicant:

1. Exhibits adequate knowledge of an instrument takeoff with instrument meteorological conditions simulated at or before reaching an altitude of 100 feet (30 meters) AGL. If accomplished in a flight simulator, visibility should be no greater than one-quarter (1/4) mile, or as specified by operator specifications.
2. Takes into account, prior to beginning the takeoff, operational factors which could affect the maneuver such as helicopter characteristics, takeoff path, surface conditions, wind, obstructions, and other related factors that could adversely affect safety.
3. Accomplishes the appropriate checklist items to ensure that the helicopter systems applicable to the instrument takeoff are operating properly.
4. Sets the applicable flight instruments to the desired setting prior to initiating the takeoff.
5. Transitions smoothly and accurately from visual meteorological conditions to actual or simulated instrument meteorological conditions.
6. Maintains the appropriate climb attitude.
7. Maintains desired heading within ±5° and desired airspeeds within ±5 knots.
8. Complies with ATC clearances and instructions issued by ATC (or the examiner simulating ATC).

C. TASK: POWERPLANT FAILURE DURING TAKEOFF

REFERENCES: FAR Part 61; AC 61-13; Pilot's Operating Handbook, RFM; DOT/FAA Takeoff Safety Training Aid.

Objective. To determine that the applicant:

1. Exhibits adequate knowledge of the procedures used during powerplant failure on takeoff, the appropriate reference airspeeds, and the specific pilot actions required.
2. Takes into account, prior to beginning the takeoff, operational factors which could affect the maneuver such as helicopter characteristics, takeoff path, surface conditions, wind, obstructions, and other related factors that could adversely affect safety.

3. Maintains the helicopter aligned with the runway heading or takeoff path appropriate for climb performance and terrain clearance when powerplant failure occurs.

4. Single-Engine Helicopters: Establishes a power-off descent approximately straight-ahead, if the powerplant failure occurs after becoming airborne. The failure of the powerplant should be simulated during a normal takeoff (no lower than 500 feet or 150 meters AGL).

5. Multiengine Helicopters: Continues the takeoff if the powerplant failure occurs at a point where the helicopter can continue to a specified airspeed and altitude at the end of the runway commensurate with the helicopter's performance capabilities and operating limitations. The failure of one powerplant should be simulated during a normal takeoff:

   a. At an appropriate airspeed that will allow continued climb performance in forward flight; or

   b. At an appropriate airspeed that is 50 percent of normal cruise speed, if there is no published single-engine airspeed for that type helicopter.

6. Maintains (in a multiengine helicopter), after a simulated powerplant failure and after a climb has been established, the desired heading within ±5° and desired airspeed within ±5 knots.

D. TASK: REJECTED TAKEOFF

REFERENCES: FAR Part 61; AC 61-13; Pilot's Operating Handbook, RFM; DOT/FAA Takeoff Safety Training Aid.

Objective. To determine that the applicant understands when to reject or continue the takeoff:

1. Exhibits adequate knowledge of the technique and procedure for accomplishing a rejected takeoff after powerplant/system(s) failure/warnings, including related safety factors.

2. Takes into account, prior to beginning the takeoff, operational factors which could affect the maneuver such as helicopter characteristics, takeoff path, surface conditions, wind, obstructions, and other related factors that could adversely affect safety.

3.   Aligns the helicopter on the runway centerline or takeoff path.
4.   Performs all required pretakeoff checks as required by the appropriate checklist items.
5.   Increases power smoothly and positively, if appropriate to the helicopter, to a predetermined value based on existing conditions.
6.   Maintains directional control on the runway heading or takeoff path.
7.   Aborts the takeoff if, in a single-engine helicopter the powerplant (or other) failure occurs prior to becoming airborne, or in a multiengine helicopter, the powerplant (or other) failure occurs at a point during the takeoff where the abort procedure can be initiated and the helicopter can be safely landed and stopped.
8.   Reduces the power smoothly and promptly, if appropriate to the helicopter, when powerplant failure is simulated. In a wheeled helicopter, the failure will be simulated at a reasonable airspeed determined after giving due consideration to the helicopter's characteristics, Height Velocity Diagram, length of landing area, surface conditions, wind direction and velocity, and any other factors that may adversely affect safety.
9.   Maintains positive control, and accomplishes the appropriate powerplant failure procedures as recommended by the appropriate checklist.

E. TASK: INSTRUMENT DEPARTURE

REFERENCES: FAR Part 61; AC 61-27; Pilot's Operating Handbook, RFM, AIM.

Objective. To determine that the applicant:

1.   Exhibits adequate knowledge of SIDs, En Route Low/High Altitude Charts, STARs and related pilot/controller responsibilities.
2.   Uses the current and appropriate navigation publications for the proposed flight.
3.   Selects and uses the appropriate communications frequencies, and selects and identifies the navigation aids associated with the proposed flight.
4.   Performs the appropriate checklist items.
5.   Establishes communications with ATC, using proper phraseology.
6.   Complies, in a timely manner, with all instructions and airspace restrictions.

7. Exhibits adequate knowledge of two-way radio communications failure procedures.

8. Intercepts, in a timely manner, all courses, radials, and bearings appropriate to the procedure, route, clearance, or as directed by the examiner.

9. Maintains the appropriate airspeed within ±10 knots, headings within ±10°, altitude within ±100 feet (30 meters); and accurately tracks a course, radial, or bearing.

10. Conducts the departure phase to a point where, in the opinion of the examiner, the transition to the en route environment is complete.

IV. INFLIGHT MANEUVERS

A. TASK: STEEP TURNS

REFERENCES: FAR Part 61; AC 61-27; FSB Report; Pilot's Operating Handbook, RFM.

Objective. To determine that the applicant:

1. Exhibits adequate knowledge of steep turns (if applicable to helicopter) and the factors associated with performance; and, if applicable, angle of bank, and pitch and power requirements.
2. Selects an altitude recommended by the manufacturer, training syllabus, or other training directive, but in no case lower than 3,000 feet (900 meters) AGL.
3. Establishes the recommended entry airspeed.
4. Rolls into a coordinated turn of 180° or 360° with a bank as appropriate, not to exceed 30°. Maintains the bank angle within ±5° while in smooth, stabilized flight.
5. Applies smooth coordinated pitch, bank, and power to maintain the specified altitude within ±100 feet (30 meters) and the desired airspeed within ±10 knots.
6. Rolls out of the turn (at approximately the same rate as used to roll into the turn) within ±10° of the entry or specified heading, stabilizes the helicopter in a straight-and-level attitude or, at the discretion of the examiner, reverses the direction of turn and repeats the maneuver in the opposite direction.
7. Avoids any indication of abnormal flight attitude, or exceeding any structural, rotor, or operating limitation during any part of the maneuver.

B. TASK: POWERPLANT FAILURE — MULTIENGINE HELICOPTER

REFERENCES: FAR Part 61; Pilot's Operating Handbook, RFM.

NOTE: Engine shutdown should be performed only under conditions, and at such altitudes (no lower than 3,000 feet (900 meters) AGL) and in a position where a safe landing can be made on an established airport, and can be accomplished in the event difficulty is encountered re-starting. At an altitude lower than 3,000 feet (900 meters) AGL, simulated engine failure will be performed by setting the powerplant controls to simulate zero-thrust. In the event the engine cannot be air started during the test, it should be treated as an emergency.

When authorized and conducted in a flight simulator, shutdown may be performed in conjunction with any procedure or maneuver and at locations and altitudes at the discretion of the examiner

Objective. To determine that the applicant:

1. Exhibits adequate knowledge of the flight characteristics and controllability associated with maneuvering with powerplant(s) inoperative (as appropriate to the helicopter).
2. Sets powerplant controls, correctly identifies and verifies the inoperative powerplant(s) after the simulated failure.
3. Maintains positive helicopter control.
4. Determines the reason for the powerplant(s) failure.
5. Follows the prescribed helicopter checklist, and verifies the procedures for securing the inoperative powerplant(s). Determines if a restart is a viable option.
6. Maintains the operating powerplant(s) within acceptable operating limits.
7. Maintains desired altitude within ±100 feet (30 meters), when a constant altitude is specified and is within the capability of the helicopter.
8. Maintains the desired airspeed within ±10 knots.
9. Maintains the desired heading within ±10° of the specified heading.
10. Demonstrates proper powerplant restart procedures in accordance with FAA approved procedure/checklist or the manufacturer's recommended procedures and pertinent checklist items.

C. TASK: POWERPLANT FAILURE — SINGLE–ENGINE HELICOPTER

REFERENCES: FAR Part 61; AC 61-13; Pilot's Operating Handbook, RFM.

NOTE: No simulated powerplant failure shall be given by the examiner in a helicopter when an actual touchdown could not be safely completed should it become necessary, nor when an autorotative descent might constitute a violation of the FARs. The examiner shall direct the applicant to terminate this TASK in a power recovery at an altitude high enough to assure that a safe touchdown could be accomplished in the event an actual powerplant failure should occur during recovery procedures.

Objective. To determine that the applicant:

1. Exhibits adequate knowledge of the flight characteristics, approach and forced (emergency) landing procedures, and related procedures to use in the event of a powerplant failure (as appropriate to the helicopter).
2. Enters autorotation promptly when the examiner simulates a powerplant failure by —

   a. lowering the collective as necessary to maintain rotor RPM within acceptable limits,
   b. establishing the recommended autorotation airspeed.
   c. maintaining proper longitudinal trim.

3. Selects a suitable airport or landing area which is within the performance capability of the helicopter.
4. Establishes a proper flight pattern to the selected airport or landing area, taking into account altitude, wind, terrain, obstructions, and other pertinent operational factors.
5. Determines the cause for the simulated powerplant failure (if altitude permits) and if a restart is a viable option.
6. Performs the emergency memory checklist items appropriate to the helicopter.
7. Maintains positive helicopter control throughout the maneuver.
8. Uses helicopter configuration devices (such as landing gear) in a manner recommended by the manufacturer and/or approved by the FAA.
9. Terminates the autorotation by performing a power recovery as briefed by the examiner, prior to the flight.

D. TASK: RECOVERY FROM UNUSUAL ATTITUDES

REFERENCES: FAR Part 61; AC 61-27; Pilot's Operating Handbook, Flight Manual.

Objective. To determine that the applicant:

1. Exhibits adequate knowledge of recovery from unusual attitudes.
2. Recovers from both nose-high and nose-low unusual attitudes, using proper pitch, bank, and power techniques.

E. TASK: SETTLING-WITH-POWER

REFERENCES: FAR Part 61; AC 61-13; Pilot's Operating Handbook, Flight Manual.

Objective. To determine that the applicant:

1. Exhibits adequate knowledge of the conditions which contribute to, and may result in, "settling-with-power."
2. Describes the relationship of gross weight, RPM, and density altitude to the severity of the vertical rate of descent.
3. At an altitude above 1,500 feet (450 meters) AGL, demonstrates entry into "settling-with-power," using the recommended procedures in the correct sequence.
4. Recovers immediately at the first indication of "settling-with-power," using the recommended procedures in the correct sequence.
5. Demonstrates smooth, positive helicopter control and prompt recovery techniques.

V. INSTRUMENT PROCEDURES

NOTE: If TASK D, Nonprecision Instrument Approach, is performed in a training device (other than an FTD or flight simulator) and the applicant has completed an approved training course for the helicopter type involved, not more than one (1) of the required instrument procedures may be observed by a person qualified to act as an instructor or check airman under that approved training program. The instrument approach is considered to begin when the helicopter is over the initial approach fix for the procedure being used and ends when the helicopter touches down on the runway or landing area, or when transition to a missed approach configuration is completed. Instrument conditions need not be simulated below the minimum altitude for the approach being accomplished.

A. TASK: INSTRUMENT ARRIVAL

REFERENCES: FAR Part 61; Pilot's Operating Handbook, RFM, AIM; En Route Low/High Altitude Charts, Profile Descent Charts, STARs, Instrument Approach Procedure Charts.

Objective. To determine that the applicant:

1. Exhibits adequate knowledge of En Route Low and High Altitude Charts, STARs, Instrument Approach Charts, and related pilot and controller responsibilities.
2. Uses the current and appropriate navigation publications for the proposed flight.
3. Selects, and correctly identifies, the appropriate navigation frequencies and facilities associated with the area arrival.
4. Performs the helicopter checklist items appropriate to the area arrival.
5. Establishes communications with ATC, using proper phraseology.
6. Complies, in a timely manner, with all ATC clearances, instructions, and restrictions.
7. Exhibits adequate knowledge of two-way communications failure procedures.
8. Intercepts, in a timely manner, all courses, radials, and bearings appropriate to the procedure, route, ATC clearance, or as directed by the examiner.
9. Adheres to airspeed restrictions and adjustments required by regulations, ATC, the RFM, or the examiner.
10. Establishes, where appropriate, a rate of descent consistent with the helicopter operating characteristics and safety.

11.   Maintains the appropriate airspeed/V-speed within ±10 knots; heading ±10°; altitude within ±100 feet (30 meters); and accurately tracks radials, courses, and bearings.

12.   Complies with the provisions of the Profile Descent, STAR, and other arrival procedures, as appropriate.

B. TASK: HOLDING

REFERENCES: FAR Part 61; Pilot's Operating Handbook, RFM, AIM; En Route Low/High Altitude Charts, STARs, Instrument Approach Procedure Charts.

Objective. To determine that the applicant:

1.   Exhibits adequate knowledge of holding procedures for standard and non-standard, published and non-published holding patterns. If appropriate, demonstrates adequate knowledge of holding endurance, including, but not necessarily limited to, fuel on board, fuel flow while holding, fuel required to alternate, etc.

2.   Changes to the recommended holding airspeed appropriate for the helicopter and holding altitude, so as to cross the holding fix at or below maximum holding airspeed.

3.   Recognizes arrival at the clearance limit or holding fix.

4.   Remains within protected airspace.

5.   Complies with ATC reporting requirements.

6.   Uses the proper timing criteria required by the holding altitude and ATC or examiner's instructions.

7.   Complies with the holding pattern leg length when a DME distance is specified.

8.   Arrives over the holding fix as close as possible to the "expect further clearance" time.

9.   Maintains the appropriate airspeed/V-speed within ±10 knots, altitude within ±100 feet (30 meters), headings within ±10°; and accurately tracks radials, courses, and bearings.

C. TASK: PRECISION INSTRUMENT APPROACHES

REFERENCES: FAR Part 61; AC 61-27; Pilot's Operating Handbook, RFM, AIM; Instrument Approach Procedure Charts.

NOTE: Two precision approaches must be accomplished.

For a multiengine helicopter, at least one manually controlled precision approach must be accomplished with a simulated failure of one powerplant. The simulated powerplant failure should occur before initiating the final approach segment and must continue to touchdown or throughout the missed approach procedure. As the markings on localizer/glide slope indicators vary, a one-quarter scale deflection of either the localizer, or glide slope indicator is when it is displaced one-fourth of the distance that it may be deflected from the on glide slope or on localizer position.

Objective. To determine that the applicant:

1. Exhibits adequate knowledge of the precision instrument approach procedures with all engines operating, and with one engine inoperative.
2. Establishes two-way communications with ATC as appropriate to the phase of flight or approach segment and uses the proper communications phraseology and techniques.
3. Accomplishes the appropriate precision instrument approach procedure as selected by the examiner.
4. Complies, in a timely manner, with all clearances, instructions, and procedures.
5. Advises ATC anytime the helicopter is unable to comply with a clearance.
6. Establishes the appropriate helicopter configuration and airspeed/V-speed considering turbulence, wind shear, microburst conditions, or other meteorological and operating conditions.
7. Completes the helicopter checklist items appropriate to the phase of flight or approach segment.
8. Prior to beginning the final approach segment, maintains the desired altitude ±100 feet (30 meters), the desired airspeed within ±10 knots, the desired heading within ±5°; and accurately tracks radials, courses, and bearings.
9. Selects, tunes, identifies, and monitors the operational status of ground and helicopter navigation equipment used for the approach.

10. Applies the necessary adjustments to the published Decision Height and visibility criteria for the helicopter approach category as required, such as —

   a. FDC and Class II NOTAMs.
   b. inoperative helicopter and ground navigation equipment.
   c. inoperative visual aids associated with the landing environment.
   d. National Weather Service (NWS) reporting factors and criteria.

11. Establishes a predetermined rate of descent at the point where the electronic glide slope begins which approximates that required for the helicopter to follow the glide slope.
12. Maintains a stabilized final approach, arriving at Decision Height with no more than one-quarter scale deflection of the localizer, or the glide slope indicators and the airspeed/V-speed within ±5 knots of that desired.
13. Avoids descent below the Decision Height before initiating a missed approach procedure or transitioning to a landing.
14. Initiates immediately the missed approach procedure, when at the Decision Height, and the required visual references for the runway or intended landing area are not distinctly visible and identifiable.
15. Transitions to a normal landing approach only when the helicopter is in a position from which a descent to a landing on the runway or intended landing area can be made at a normal rate of descent using normal maneuvering.

D. TASK: NONPRECISION INSTRUMENT APPROACHES

REFERENCES: FAR Part 61; AC 61-27; Pilot's Operating Handbook, RFM, AIM; Instrument Approach Procedure Charts.

NOTE: The applicant must accomplish at least two nonprecision approaches. The examiner will select instrument nonprecision approach procedures that are representative of that which the applicant is likely to use. The second nonprecision approach procedure will utilize a navigational aid other than the one used for the first approach.

Objective. To determine that the applicant:

1. Exhibits adequate knowledge of nonprecision approach procedures representative of those the applicant is likely to use.
2. Establishes two-way communications with ATC as appropriate to the phase of flight or approach segment and uses proper communications phraseology and techniques.
3. Accomplishes the nonprecision instrument approach procedures selected by the examiner.
4. Complies with all clearances issued by ATC.
5. Advises ATC or the examiner any time the helicopter is unable to comply with a clearance.
6. Establishes the appropriate helicopter configuration and airspeed, and completes all applicable checklist items.
7. Maintains, prior to beginning the final approach segment, the desired altitude ±100 feet (30 meters), the desired airspeed ±10 knots, the desired heading ±5°; and accurately tracks radials, courses, and bearings.
8. Selects, tunes, identifies, and monitors the operational status of ground and helicopter navigation equipment used for the approach.
9. Applies the necessary adjustments to the published Minimum Descent Altitude and visibility criteria for the helicopter approach category when required, such as —

    a. Notices to Airmen, including Flight Data Center Procedural NOTAMs.
    b. inoperative helicopter and ground navigation equipment.
    c. inoperative visual aids associated with the landing environment.
    d. National Weather Service (NWS) reporting factors and criteria.

10. Establishes a rate of descent that will ensure arrival at the Minimum Descent Altitude with the helicopter in a position from which a descent to a landing on the intended runway or landing area can be made at a normal rate using normal maneuvering.
11. Allows, while on the final approach segment, not more than quarter-scale deflection of the Course Deviation Indicator (CDI) or ±5° in the case of the RMI or bearing pointer, and maintains airspeed within ±5 knots of that desired.
12. Maintains the Minimum Descent Altitude, when reached, within −0, +50 feet (−0, +15 meters) to the missed approach point.

13. Executes the missed approach procedure if the required visual references for the intended runway are not distinctly visible and identifiable at the missed approach point.

14. Executes a normal landing from a straight-in approach.

E. TASK: MISSED APPROACH

REFERENCES: FAR Part 61; AC 61-27; Pilot's Operating Handbook, RFM, AIM; Instrument Approach Procedure Charts.

NOTE: The applicant must be required to perform at least two missed approach procedures with at least one missed approach from a precision approach (ILS, MLS, or GPS). A complete approved missed approach procedure must be accomplished at least once and a simulated powerplant failure (in a multiengine helicopter) will be required during one of the missed approaches.

Going below the MDA or DH, as appropriate, prior to the initiation of the missed approach procedure shall be considered unsatisfactory performance, except in those instances where the required visual references for the runway or intended landing area are distinctly visible and identifiable at the MDA or DH.

Objective. To determine that the applicant:

1. Exhibits adequate knowledge of missed approach procedures associated with standard instrument approaches.

2. Initiates the missed approach procedure promptly by the timely application of power, establishes the proper climb attitude, and reduces drag in accordance with the approved procedures.

3. Reports to ATC, beginning the missed approach procedure.

4. Complies with the appropriate missed approach procedure or ATC clearance.

5. Advises ATC any time the helicopter is unable to comply with a clearance.

6. Follows the recommended helicopter checklist items appropriate to the go-around procedure for the helicopter used.

7. Requests clearance, if appropriate, to the alternate airport, another approach, a holding fix, or as directed by the examiner.

8. Maintains the desired altitudes ±100 feet (30 meters), airspeed ±5 knots, heading ±5°; and accurately tracks courses, radials, and bearings.

VI. LANDINGS AND APPROACHES TO LANDINGS

NOTE: Notwithstanding the authorizations for the combining of maneuvers and for the waiver of maneuvers, the applicant must make at least four (4) landings to a hover or to the ground. These landings must include the types listed in this AREA OF OPERATION; however, more than one type may be combined where appropriate (i.e., crosswind and landing from a precision approach or landing with simulated powerplant failure, etc.).

A. TASK: NORMAL AND CROSSWIND APPROACHES AND LANDINGS

REFERENCES: FAR Part 61; AC 61-13; Pilot's Operating Handbook, RFM.

Objective. To determine that the applicant:

1. Exhibits adequate knowledge of normal and crosswind approaches and landings including recommended approach angles, airspeeds, V-speeds, configurations, performance limitations, wake turbulence, and safety factors (as appropriate to the helicopter).
2. Establishes the approach and landing configuration appropriate for the runway or designated landing area and meteorological conditions, and adjusts the powerplant controls as required.
3. Maintains a ground track that ensures the desired traffic pattern will be flown, taking into account any obstructions and ATC or examiner instructions.
4. Verifies existing wind conditions, makes proper correction for drift, and maintains a precise ground track.
5. Maintains a normal approach angle and recommended airspeed and a normal rate of closure to the point of transition to a hover or touchdown.
6. Terminates the approach in a smooth transition to a hover or to a touchdown within 2 feet (.6 meter) of the designated point. (If a hover termination is specified, it will be within ±2 feet (.6 meter) of recommended hovering altitude.)
7. Completes the applicable after-landing checklist items in a timely manner and as recommended by the manufacturer.

B. TASK: APPROACH AND LANDING WITH SIMULATED POWERPLANT FAILURE — MULTIENGINE HELICOPTER

REFERENCES: FAR Part 61; AC 61-13; Pilot's Operating Handbook, RFM.

NOTE: In a multiengine helicopter maneuvering to a landing, the applicant should follow a procedure that simulates the loss of one powerplant.

Objective. To determine that the applicant:

1. Exhibits adequate knowledge of maneuvering to a landing with a powerplant inoperative, including the controllability factors associated with maneuvering, and the applicable emergency procedures.
2. Proceeds toward the nearest suitable airport or landing area.
3. Maintains, prior to beginning the final approach segment, the desired altitude ±100 feet (30 meters), the desired airspeed ±10 knots, the desired heading ±5°; and accurately tracks courses, radials, and bearings.
4. Establishes the approach and landing configuration appropriate for the runway or landing area, and meteorological conditions; and adjusts the powerplant controls as required.
5. Maintains a normal approach angle and recommended airspeed to the point of transition to touchdown.
6. Terminates the approach in a smooth transition to touchdown.
7. Completes the after-landing checklist items in a timely manner, after clearing the runway, and as recommended by the manufacturer.

C. TASK: REJECTED LANDING

REFERENCES: FAR Part 61; AC 61-13; Pilot's Operating Handbook, RFM; FSB Report.

NOTE: The maneuver may be combined with instrument or missed approach procedures, but instrument conditions need not be simulated below 100 feet (30 meters) above the runway or landing area. This maneuver should be initiated approximately 50 feet (15 meters) above the runway and approximately over the runway threshold or as recommended by the FSB Report.

Objective. To determine that the applicant:

1. Exhibits adequate knowledge of a rejected landing procedure, including the conditions that dictate a rejected landing, the importance of a timely decision, the recommended airspeed/V-speeds, and also the applicable "clean-up" procedure.
2. Makes a timely decision to reject the landing for actual or simulated circumstances.
3. Applies the appropriate power setting for the flight condition and establishes a pitch attitude necessary to obtain the desired performance.
4. Adjusts helicopter configuration and retracts the landing gear, if appropriate, in the correct sequence and at a safe altitude, establishes a positive rate of climb and the appropriate airspeed/V-speed within ±5 knots.
5. Trims the helicopter as necessary, and maintains the proper ground track during the rejected landing procedure.
6. Accomplishes the appropriate checklist items in a timely manner in accordance with approved procedures.

VII. NORMAL AND ABNORMAL PROCEDURES

REFERENCES: FAR Part 61; AC 61-13; Pilot's Operating Handbook, RFM.

Objective. To determine that the applicant:

1. Possesses adequate knowledge of the normal and abnormal procedures of the systems, subsystems, and devices relative to the helicopter type (as may be determined by the examiner).

2. Demonstrates the proper use of the helicopter's systems, subsystems, and devices (as may be determined by the examiner) appropriate to the helicopter, such as —

   a. powerplant.
   b. fuel system.
   c. electrical system.
   d. hydraulic system.
   e. environmental system.
   f. fire detection and extinguishing systems.
   g. navigation and avionics systems.
   h. automatic flight control system, electronic flight instrument system, and related subsystems.
   i. flight control systems.
   j. anti-ice and deice systems.
   k. helicopter and personal emergency equipment.
   l. loss of tail rotor effectiveness.
   m. other systems, subsystems, and devices specific to the type helicopter.

VIII. EMERGENCY PROCEDURES

REFERENCES: FAR Part 61; Pilot's Operating Handbook, RFM.

Objective. To determine that the applicant:

1. Possesses adequate knowledge of the emergency procedures (as may be determined by the examiner) relating to the particular helicopter type.
2. Demonstrates the proper emergency procedures (as must be determined by the examiner) relating to the particular helicopter type, including —

   a. inflight fire and smoke removal
   b. emergency descent.
   c. autorotation, with a power recovery.
   d. ditching.
   e. emergency evacuation.

3. Demonstrates the proper procedure for any other emergency outlined (as must be determined by the examiner) in the appropriate approved helicopter RFM.

IX. POSTFLIGHT PROCEDURES

A. TASK: AFTER-LANDING PROCEDURES

REFERENCES: Pilot's Operating Handbook, RFM.

Objective. To determine that the applicant:

1. Exhibits adequate knowledge of safe after-landing/taxi procedures (as appropriate to the helicopter).
2. Demonstrates proficiency by maintaining correct and positive helicopter control. This includes hovering taxi, air taxiing; and in helicopters with wheels, includes ground taxiing. In helicopters equipped with float devices, this includes water taxiing, approaching a buoy, and docking.
3. Maintains proper spacing on other helicopter, obstructions, and persons.
4. Accomplishes the applicable checklist items and performs the recommended procedures.
5. Maintains the desired track and speed.
6. Complies with instructions issued by ATC (or the examiner simulating ATC).
7. Observes runway hold lines, localizer and glide slope critical areas, and other surface control markings and lighting.
8. Maintains constant vigilance and control of the helicopter during the taxi operation.

B. TASK: PARKING AND SECURING

REFERENCES: Pilot's Operating Handbook, RFM.

Objective. To determine that the applicant:

1. Exhibits adequate knowledge of the parking and the securing helicopter procedures.
2. Applicant has adequate knowledge of the helicopter forms/logs to record the flight time/discrepancies.

Appendix 2 - Helicopters

TASK VS. SIMULATION DEVICE CREDIT

Examiners conducting the Airline Transport Pilot and/or Type Rating Practical Tests with simulation devices should consult appropriate documentation to ensure that the device has been approved for training and checking the TASKS in question. The documentation for each device should reflect that the following activities have occurred:

1. The device must be evaluated, determined to meet the appropriate standards, and assigned the appropriate qualification level by the National Simulator Program Manager. The device must continue to meet qualification standards through continuing evaluations as outlined in the appropriate advisory circular (AC). For helicopter simulators, AC 120-63 (as amended), Helicopter Simulator Qualification, will be used.

2. The FAA must approve the device for specific TASKS.

3. The device must continue to support the level of student or applicant performance required by this PTS.

NOTE: Users of the following chart are cautioned that use of the chart alone is incomplete. The description and objective of each task as listed in the body of the PTS, including all notes, must also be incorporated for accurate simulation device use.

USE OF CHART

X Creditable.

X1 Creditable only if accomplished in conjunction with a running takeoff or running landing, as appropriate.

NOTE:
1. The helicopter may be used for all tasks.
2. Level C simulators may be used as indicated only if the applicant meets established pre-requisite experience requirements.
3. Level A helicopter simulator standards have not been defined.
4. Helicopter flight training devices have not been defined.

FLIGHT TASK
Areas of Operation: Section Two

	1	2	3	4	5	6	7	A	B	C	D
I. Preflight Procedures											
A. Preflight Inspection (Cockpit Only)									X	X	X
II. Ground Operation											
A. Powerplant Start									X	X	X
B1. Taxi - Ground										X	X
B2. Taxi - Hover										X	X
C. Pretakeoff Checks									X	X	X
III. Takeoff and Departure Maneuvers											
A. Normal and Crosswind Takeoff									X1	X	X
B. Instrument Takeoff									X1	X	X
C. Powerplant Failure During Takeoff									X1	X	X
D. Rejected Takeoff									X1	X	X
E. Instrument Departure									X	X	X
IV. Inflight Maneuvers											
A. Steep Turns									X	X	X
B. Powerplant Failure - Multiengine Helicopters									X	X	X
C. Powerplant Failure - Single-Engine Helicopters									X	X	X
D. Recovery From Unusual Attitudes									X	X	X
E. Settling-With-Power										X	X

LEVEL of SIMULATION DEVICE

	1	2	3	4	5	6	7	A	B	C	D
V. Instrument Procedures											
A. Instrument Arrival									X	X	X
B. Holding									X	X	X
C1. Precision Instrument Approach (Normal)									X	X	X
C2. Precision Inst. Approach (Manual/Pwrplnt Fail.)										X	X
D. Nonprecision Instrument Approaches										X	X
E1. Missed Approach (Normal)									X	X	X
E2. Missed Approach (Powerplant Failure)									X	X	X
VI. Landings and Approaches to Landings											
A. Normal and Crosswind Approaches and Landings										X1	X
B. Landing with Powerplant Failure									X1	X	X
C. Rejected Landing									X	X	X
VII. Normal and Abnormal Procedures (*1)											
A. Powerplant									X	X	X
B. Fuel System									X	X	X
C. Electrical System									X	X	X
D. Hydraulic System									X	X	X
E. Environmental System(s)									X	X	X

(*1) Evaluation of normal and abnormal procedures can usually be accomplished in conjunction with other events and does not normally require a specific event to test the applicant's use of the aircraft systems and devices. An applicant's performance must be evaluated on the maintenance of helicopter control, the ability to recognize and analyze abnormal indications, and the ability to apply corrective procedures in a timely manner.

VII. Normal and Abnormal Procedures (Cont.) (*1)

	LEVEL of SIMULATION DEVICE										
	1	2	3	4	5	6	7	A	B	C	D
F. Fire Detection and Extinguisher Systems									X	X	X
G. Navigation and Aviation Systems									X	X	X
H. Automatic Flight Control System, Electronic Flight Instrument System, and Related Subsys.										X	X
I. Flight Control Systems									X	X	X
J. Anti-ice and Deice Systems									X	X	X
K. Aircraft and Personal Emergency Equipment										X	X
L. Loss of Tail Rotor Effectiveness										X	X
M. Others, as determined by make, model, or series											

VII. Emergency Procedures

	1	2	3	4	5	6	7	A	B	C	D
A. Emergency Descent									X	X	X
B. Inflight Fire and Smoke Removal									X	X	X
C. Emergency Evacuation										X	X
D. Ditching										X	X
E. Autorotative Landing										X	X

IX. Postflight Procedures

	1	2	3	4	5	6	7	A	B	C	D
A. After Landing									X	X	X
B. Parking and Securing									X	X	X

(*1) Evaluation of normal and abnormal procedures can usually be accomplished in conjunction with other events and devices. An applicant's performance must be evaluated on the maintenance of specific event to test the applicant's use of the aircraft systems and devices. An applicant's performance must be evaluated on the maintenance of helicopter control, the ability to recognize and analyze abnormal indications, and the ability to apply corrective procedures in a timely manner.

U.S. Department
of Transportation
**Federal Aviation
Administration**

FAA-S-8081-10A
Effective June 1995

Aircraft Dispatcher

Practical Test
Standards

Flight Standards Service
Washington, DC 20591

Reprinted by Aviation Supplies & Academics, Inc.
Newcastle, WA 98059-3153

ii

NOTE

Material in FAA-S-8081-10A, Aircraft Dispatcher Practical Test Standards, becomes effective June 1, 1995. All previous editions of this book become obsolete as of this date.

iv

FOREWORD

The Aircraft Dispatcher Practical Test Standards (PTS) book has been published by the Federal Aviation Administration (FAA) to establish the standards for the aircraft dispatcher practical test. FAA inspectors and designated examiners shall conduct practical tests in compliance with these standards. Instructors and applicants should find these standards helpful in practical test preparation.

William J. White
Deputy Director, Flight Standards Service

vi

INTRODUCTION

The Flight Standards Service of the FAA has developed this practical test book as a standard to be used by FAA inspectors and designated examiners when conducting the aircraft dispatcher practical test. Instructors are expected to use this book when preparing applicants for the practical test.

This publication sets forth the practical test requirements for the aircraft dispatcher certificate.

Information considered directive in nature is described in this practical test standard in terms such as "shall" and "must," and means that the actions are mandatory. Guidance information is described in terms such as "will," "should," or "may," and indicate actions that are desirable, permissive, or not mandatory and provide for flexibility.

The FAA gratefully acknowledges the valuable assistance provided by organizations and individuals who have contributed their time and talent in the development of the practical test standards.

This publication may be purchased from the Superintendent of Documents, U.S. Government Printing Office, Washington, DC 20402.

Comments regarding this publication should be sent to:

U.S. Department of Transportation
Federal Aviation Administration
Flight Standards Service
Operations Support Branch, AFS-630
P.O. Box 25082
Oklahoma City, OK 73125

PRACTICAL TEST STANDARD CONCEPT

Federal Aviation Regulations (FARs) specify the areas in which knowledge and skill must be demonstrated by the applicant before the issuance of an aircraft dispatcher certificate. The FARs provide the flexibility to permit the FAA to publish practical test standards containing specific TASKS in which competency must be demonstrated by the applicant before the issuance of an aircraft dispatcher certificate. The FAA will revise this book whenever it is determined that changes are needed in the interest of safety. Adherence to provisions of the FARs and the practical test standards is mandatory for the evaluation of aircraft dispatcher applicants.

EXAMINER[1] RESPONSIBILITY

The examiner who conducts the practical test is responsible for determining that the applicant meets the standards outlined in the Objective of each TASK within the appropriate practical test standard. The examiner shall meet this responsibility by accomplishing an action that is appropriate for each TASK. For each TASK that involves "knowledge only" elements, the examiner shall orally quiz the applicant on those elements. For each TASK that involves both "knowledge and skill" elements, the examiner shall orally quiz the applicant regarding knowledge elements and ask the applicant to perform the required skill elements. The examiner shall determine that the applicant's knowledge and skill meet the Objective in all required TASKS. Oral questioning may be used at any time during the practical test.

NOTE: Where appropriate, the applicant should be allowed to use reference material.

CREW RESOURCE MANAGEMENT (CRM)

CRM "... refers to the effective use of ALL available resources; human resources, hardware, and information." Human resources "... includes all other groups routinely working with the cockpit crew (or pilot) who are involved in decisions that are required to operate a flight safely. These groups include, but are not limited to: dispatchers, cabin crewmembers, maintenance personnel, and air traffic controllers." CRM is not a single TASK. CRM is a set of skill competencies which must be evident in all TASKS in this PTS. CRM competencies, grouped into three clusters of observable behavior are:

1. COMMUNICATIONS PROCESSES AND DECISIONS

   a. Briefing
   b. Inquiry/Advocacy/Assertiveness
   c. Self-Critique
   d. Communication with available personnel resources
   e. Decision making

---

[1] The word "examiner" is used throughout this standard to denote either the FAA inspector or FAA designated examiner who conducts the official practical test.

2. BUILDING AND MAINTENANCE OF A FLIGHT TEAM

   a. Leadership/Followership
   b. Interpersonal Relationships

3. WORKLOAD MANAGEMENT AND SITUATIONAL AWARENESS

   a. Preparation/Planning
   b. Workload Distribution
   c. Distraction Avoidance

Examiners are required to exercise proper CRM competencies in conducting tests as well as expecting the same from applicants.

CRM deficiencies almost always contribute to the unsatisfactory performance of a TASK. Therefore, the competencies provide an extremely valuable vocabulary for debriefing. For debriefing purposes, an amplified list of these competencies expressed as behavioral markers, may be found in appendix 1 of AC 120-51A. These markers consider the use of various levels of automation in flight management systems.

AIRCRAFT DISPATCHER PRACTICAL TEST STANDARD DESCRIPTION

The AREAS OF OPERATION are subjects in which an aircraft dispatcher must have knowledge and demonstrate skill. They begin with the preparation for the flight and end with emergency procedures. The examiner, however, may conduct the practical test in any sequence that results in a complete and efficient test.

The Objective lists the important elements that must be satisfactorily performed to demonstrate competency in a TASK. The Objective includes:

1. specifically what the applicant should be able to do;
2. the conditions under which the TASK is to be performed; and
3. the minimum acceptable standards of performance.

The REFERENCE identifies the publication(s) that describes or refers to the TASK. Descriptions of TASKS are not included in the aircraft dispatcher standards because this information can be found in references listed for each TASK. Publications other than those listed may be used as references if their content conveys substantially the same meaning as the referenced publications.

References upon which this practical test book is based include:

FAR Part 65	Certification: Airmen Other Than Flight Crewmembers
FAR Part 121	Certification and Operations: Domestic, Flag, and Supplemental Air Carriers and Commercial Operators of Large Aircraft
HMR 175	Hazardous Materials Regulations
NTSB PART 830	Notification and Reporting of Aircraft Accidents and Incidents
AC 00-6	Aviation Weather
AC 00-45	Aviation Weather Services
AC 61-27	Instrument Flying Handbook
AC 120-51	Crew Resource Management Training
DOT\FAA\RD	Crew Resource Management: An October 1992 Introductory Handbook
AIM	Airman's Information Manual
SIDs	Standard Instrument Departures
STARs	Standard Terminal Arrivals
AFD	Airport/Facility Directory
FDC NOTAMs	National Flight Data Center Notices to Airmen
IAP	Instrument Approach Procedure En Route Low/High Altitude Charts, Pertinent Pilot Operating Handbooks and FAA-Approved Flight Manuals, Operations Specifications, Minimum Equipment List (MEL) and Configuration Deviation List (CDL)

NOTE: The latest revision of the references should be used.

USE OF THE PRACTICAL TEST STANDARDS BOOK

This practical test book contains only one practical test standard. When using the practical test book, the examiner must evaluate the applicant's knowledge and skill in sufficient depth to determine that the standards of performance listed for all TASKS are met.

All TASKS in this practical test standard are required for the issuance of an aircraft dispatcher certificate. However, when a particular ELEMENT is not appropriate to the aircraft, its equipment, or operational capability, that ELEMENT, at the discretion of the examiner, may be omitted. It is not intended that the examiner follow the precise order in which AREAS OF OPERATION and TASKS appear in the test book. The examiner may change the sequence or combine TASKS with similar Objectives to conserve time. The examiner shall develop a plan of action that includes the order and combination of TASKS to be demonstrated by the applicant in a manner that results in an efficient and valid test.

TASKS with similar Objectives may be combined to conserve time; however, the Objectives of all TASKS must be demonstrated at some time during the practical test. It is of the utmost importance that the examiner accurately evaluates the applicant's ability to perform safely as an aircraft dispatcher.

The examiner shall place special emphasis upon AREAS OF OPERATION which are most critical to flight safety. One of these areas is sound judgment in decision making. Although these areas may not be shown under each TASK, they are essential to flight safety and shall receive careful evaluation throughout the practical test.

In an automated environment, the examiner must require an applicant to demonstrate manual flight planning.

AIRCRAFT DISPATCHER PRACTICAL TEST PREREQUISITES

An applicant for an aircraft dispatcher practical test is required by the FARs to:

1. have passed the appropriate aircraft dispatcher knowledge test since the beginning of the 24th month before the month in which the practical test is taken; and
2. obtained the applicable experience prescribed for the aircraft dispatcher certificate under FAR Section 65.57 and must provide documentary evidence of such experience or
3. have successfully completed an FAA-approved aircraft dispatcher training course within the past 90 days.

REQUIRED MATERIAL FOR THE PRACTICAL TEST

The examiner is responsible for supplying weather data for the test when current weather information is not available.

Materials to be supplied by the applicant are:

1. Company Aircraft Operating Manual or Flight Manual.
2. General Operations Manual and Operations Specifications.
3. En Route Low/High Altitude Charts.
4. Standard Instrument Departures.
5. Standard Terminal Arrival Routes.
6. Standard Instrument Approach Procedures Charts.
7. Flight Plan Form.
8. Load Manifest Form.
9. Dispatch Release Form.
10. Airman's Information Manual/International Information Manual.
11. Computer and Plotter.

SATISFACTORY PERFORMANCE

The ability of an applicant to perform the required TASKS is based on:

1. showing competency within the standards outlined in this test book;
2. following emergency procedures as required by the FARs and company procedures;
3. exercising good judgment; and
4. applying aeronautical knowledge.

UNSATISFACTORY PERFORMANCE

If, in the judgment of the examiner, the applicant does not meet the standards of performance of any TASK performed, the associated AREA OF OPERATION is failed and; therefore, the practical test is failed. The examiner or applicant may discontinue the test at any time after the failure of an AREA OF OPERATION makes the applicant ineligible for the certificate sought. The test shall be continued only with the consent of the applicant. If the test is either continued or discontinued, the applicant is entitled to credit for only those AREAS OF OPERATION satisfactorily performed. However, during the retest and at the discretion of the examiner, any TASK may be re-evaluated, including those previously passed.

RECORDING UNSATISFACTORY PERFORMANCE

The term "AREA OF OPERATION" is used to denote areas in which the applicant must demonstrate competency prior to being issued an aircraft dispatcher certificate. This practical test book uses the terms "AREAS OF OPERATION" and "TASK" to denote areas in which competency must be demonstrated. When a disapproval notice is issued, the examiner shall record the applicant's unsatisfactory performance in terms of AREAS OF OPERATION appropriate to the practical test conducted.

CONTENTS

I. AREA OF OPERATION: DISPATCH EXERCISE

A. TASK: FLIGHT PLANNING

REFERENCES: FAR Parts 65, 121.

NOTE: Where appropriate, questions on other AREAS OF OPERATION may be based on the assigned flight.

Objective. To determine that the applicant:

1. Exhibits adequate knowledge of flight planning by preparing a flight plan, load manifest, and dispatch release for a flight between designated points.

2. Plans the flight in accordance with regulatory requirements and company procedures, as appropriate.

B. TASK: OBTAINING WEATHER INFORMATION

REFERENCES: FAR Part 65; AC 00-6, AC 00-45; AIM.

NOTE: Where current weather reports, forecasts, or other pertinent information is not available, this information shall be simulated by the examiner in a manner which adequately measures the applicant's competence.

Objective. To determine that the applicant:

1. Exhibits adequate knowledge of aviation weather information by obtaining, reading, and analyzing the applicable items such as —

   a. weather reports and forecasts.
   b. pilot and radar reports.
   c. surface analysis charts.
   d. radar summary charts.
   e. significant weather prognostics.
   f. winds and temperatures aloft.
   g. freezing level charts.
   h. stability charts.
   i. severe weather outlook charts.
   j. constant pressure charts.

       k. constant pressure prognostics.

       l.   tables and conversion graphs.

      m. SIGMETs and AIRMETs.

      n. ATIS reports.

      o. NOTAMs/NOTAM systems.

2. Correctly analyzes the assembled weather information pertaining to the proposed route of flight and destination airport, and determines whether an alternate airport is required. If required, determine whether the selected alternate meets the requirements of the FARs and the operations specifications.

II. AREA OF OPERATION: AIRCRAFT

A. TASK: FLIGHT INSTRUMENTS

REFERENCES: FAR Part 65; AC 61-27.

Objective. To determine that the applicant exhibits adequate knowledge of the applicable aircraft flight instruments and systems, and their operating characteristics such as:

1. Altimeter.
2. Airspeed indicator.
3. Vertical-speed indicator.
4. Attitude indicator.
5. Horizontal situation indicator.
6. Magnetic compass.
7. Turn-and-slip indicator.
8. Heading indicator.

B. TASK: NAVIGATION INSTRUMENTS AND AVIONIC SYSTEMS

REFERENCES: FAR Part 65; AC 61-27; Operating Handbook, Flight Manual.

Objective. To determine that the applicant exhibits adequate knowledge of the applicable aircraft navigation instruments and avionics systems, and their operating methods such as:

1. VHF omnirange (VOR).
2. Distance measuring equipment (DME).
3. Instrument landing system (ILS)/microwave landing system (MLS).
4. Marker beacon receiver/indicators.
5. Transponder/altitude encoding.
6. Automatic direction finding (ADF).
7. Electronic flight indicating system (EFIS).
8. Long range navigation (LORAN).
9. Inertial navigation system (INS).
10. Radio area navigation (RNAV).
11. Doppler radar.
12. Autopilot and flight director.
13. Communications equipment.
14. Global positioning system (GPS).
15. Flight management system (FMS).

C. TASK: AIRCRAFT SYSTEMS

REFERENCES: FAR Part 65; Company Aircraft Operating Manual, Flight Manual.

Objective. To determine that the applicant exhibits adequate knowledge of the aircraft; its systems and components; its normal, abnormal, and emergency operating procedures; and (as appropriate to the aircraft) the use of correct terminology with regard to such items as:

1. Landing gear.
2. Powerplant/systems/components (reciprocating, turboprop, turbojet).
3. Fuel system.
4. Oil system.
5. Hydraulic system.
6. Electrical system.
7. Environmental system.
8. Ice protection.

D. TASK: AIRCRAFT HANDBOOKS, MANUALS, MINIMUM EQUIPMENT LIST, AND OPERATIONS SPECIFICATIONS

REFERENCES: FAR Parts 65, 121; Company Aircraft Operating Manual, Flight Manual; Minimum Equipment List; Operations Specifications.

Objective. To determine that the applicant exhibits adequate knowledge of the operating handbook or flight manual with regard to TASKS A, B, and C, and the minimum equipment list, and operations specifications as appropriate.

E. TASK: AIRCRAFT PERFORMANCE AND LIMITATIONS

REFERENCES: FAR Parts 65, 121; Company Aircraft Operating Manual , Flight Manual.

Objective. To determine that the applicant:

1. Exhibits adequate knowledge of performance limitations, including thorough knowledge of the adverse effects of exceeding any limitation.
2. Demonstrates proficient use of (as appropriate to the aircraft) performance charts, tables, graphs, or other data relating to such items as —

   a. accelerate-stop distance.
   b. accelerate-go distance.
   c. takeoff performance, all engines and engine(s) inoperative.
   d. climb performance, all engines and engine(s) inoperative.
   e. service ceiling, all engines and engine(s) inoperative.
   f. cruise performance.
   g. fuel consumption, range, and endurance.
   h. descent performance.
   i. go-around from rejected landing.
   j. drift down.

3. Describes (as appropriate to the aircraft) the performance airspeeds used during specific phases of flight.
4. Describes the effects of meteorological conditions upon performance characteristics and correctly applies these factors to a specific chart, table, graph, or other performance data.
5. Computes the center-of-gravity location for a specific load condition (as specified by the examiner), including adding, removing, and shifting weight.
6. Determines that the takeoff weight, landing weight, and zero fuel weight are within limits.
7. Demonstrates good planning and knowledge of procedures in applying operational factors affecting aircraft performance.

III. AREA OF OPERATION:
AIR ROUTES AND AIRPORTS

A. TASK: ROUTING

REFERENCES: FAR Parts 65, 121.

Objective. Using the appropriate en route charts, the applicant should:

1. Show the correlation and transition from one portion of the flight to another (SID to low altitude en route to high altitude enroute).
2. Describe the route over which the flight is to be dispatched including —

   a. intermediate stops.
   b. alternate airports.
   c. refueling and provisional airports.

B. TASK: USE AND INTERPRETATION OF SIDS, EN ROUTE CHARTS, STARS, AND STANDARD INSTRUMENT APPROACH PROCEDURES

REFERENCES: FAR Part 65; AIM; Airport/Facility Directory; SIDs; STARs; En Route Low/High Altitude Charts, Standard Instrument Approach Charts.

Objective. To determine that the applicant:

1. Understands and can define such items as —

   a. minimum en route altitude (MEA).
   b. minimum reception altitude (MRA).
   c. minimum obstacle clearance altitude (MOCA).
   d. minimum crossing altitude (MCA).
   e. standard instrument departure (SID).
   f. standard terminal arrival procedure (STAR).
   g. preferred routes.
   h. RNAV routes.

2. Can locate such item on SIDs and en route charts as —

    a. VOR/VORTACS.
    b. compulsory/non-compulsory reporting points.
    c. VOR changeover points.
    d. DME fix.
    e. airway intersection.
    f. symbols for MEA, MCA, and MRA.
    g. clearance limit or transition (SID).

3. Can locate and discuss the following information on the appropriate instrument approach procedures chart —

    a. field elevation.
    b. touchdown zone elevation (TDZE).
    c. aircraft approach category.
    d. decision height (DH) and/or minimum descent altitude (MDA).
    e. IFR approach/landing minimums (straight-in, circling, side-step, and radar).
    f. takeoff minimums (standard/non-standard).
    g. availability of radar service.
    h. procedure turn limitations.
    i. time/distance from final approach fix (FAF) to missed approach point (MAP).
    j. published missed approach procedure.
    k. obstructions.

C. TASK: AIRPORTS

REFERENCE: Airport/Facility Directory.

Objective. To determine that the applicant can:

1. Describe such items as the following at a specified airport —

   a. runway lengths.
   b. primary runway gradient and width.
   c. displaced thresholds.
   d. approach lighting system.
   e. availability of VASI.
   f. runway lighting system.

2. Discuss the following as it relates to the assigned dispatch —

   a. runway visual range.
   b. effect of inoperative components and visual aids on landing minimums.
   c. IFR landing minimums for the alternate airport.
   d. requirement with regard to the alternate for the departure airport.

IV. AREA OF OPERATION:
AIRMAN'S INFORMATION MANUAL

REFERENCES: FAR Part 65; AIM.

Objective. To determine that the applicant has a working knowledge of the Airman's Information Manual and is able to discuss such topics as:

1. Navigational aids.
2. Airport/air navigation lighting and marking.
3. Airspace.
4. Air traffic control.
5. Airport operations.
6. Air traffic control clearances.
7. Preflight.
8. Departure/en route/arrival procedures.

V. AREA OF OPERATION:
DISPATCH AND OPERATIONAL CONTROL

A. TASK: COMPANY OPERATIONS

REFERENCES: FAR Parts 65, 121; General Operations Manual; Operations Specifications.

Objective. To assure the applicant has knowledge of company procedures by discussing such items as:

1. Dispatch area, routes, and main terminals.
2. Approved instrument approach procedures.
3. Takeoff and landing minimums.
4. The difference in decision height as it relates to category (CAT I - CAT II - CAT III).
5. Use of minimum equipment list (MEL).
6. Configuration deviation list (CDL).
7. Air traffic flow control.
8. Redispatch.

B. TASK: REGULATORY REQUIREMENTS

REFERENCES: FAR Parts 65, 121; HMR 175.

Objective. To assure the applicant has adequate knowledge of regulations pertaining to the dispatch and operational control of a flight by discussing such items as:

1. Dispatcher responsibilities.
2. Dispatcher/pilot responsibilities.
3. Required equipment.

VI. AREA OF OPERATION: EMERGENCY PROCEDURES

A. TASK: COMPANY POLICY

REFERENCES: FAR Parts 65, 121; General Operations Manual.

Objective. To ensure the applicant has knowledge of company procedures regarding emergency situations.

B. TASK: OTHER PROCEDURES AND SERVICES

REFERENCES: FAR Parts 65, 121; NTSB Part 830; AIM.

Objective. To ensure the applicant is familiar with the following services and procedures:

1. Responsibility for declaring an emergency.
2. Required reporting of an emergency.
3. Collection and dissemination of information on overdue or missing aircraft.
4. FAA responsibility and services.
5. Means of declaring an emergency.
6. NTSB reporting requirements.